Springer Geography

The Springer Geography series seeks to publish a broad portfolio of scientific books, aiming at researchers, students, and everyone interested in geographical research. The series includes peer-reviewed monographs, edited volumes, textbooks, and conference proceedings. It covers the entire research area of geography including, but not limited to, Economic Geography, Physical Geography, Quantitative Geography, and Regional/Urban Planning.

More information about this series at http://www.springer.com/series/10180

Viacheslav I. Vasenev · Elvira Dovletyarova
Zhongqi Cheng · Riccardo Valentini
Editors

Megacities 2050: Environmental Consequences of Urbanization

Proceedings of the VI International Conference on Landscape Architecture to Support City Sustainable Development

Springer

Editors
Viacheslav I. Vasenev
RUDN University
Moscow
Russia

Elvira Dovletyarova
RUDN University
Moscow
Russia

Zhongqi Cheng
City University of New York
New York
USA

Riccardo Valentini
Euro-Mediterranean Center for Climate Change
Viterbo
Italy

ISSN 2194-315X ISSN 2194-3168 (electronic)
Springer Geography
ISBN 978-3-319-70556-9 ISBN 978-3-319-70557-6 (eBook)
https://doi.org/10.1007/978-3-319-70557-6

Library of Congress Control Number: 2017957853

Printed on acid-free paper

This Springer imprint is published by Springer Nature
The registered company is Springer International Publishing AG
The registered company address is: Gewerbestrasse 11, 6330 Cham, Switzerland

Preface

This edited volume contains a selection of refereed and revised papers originally presented at the Sixth International Conference on Landscape Architecture to Support City Sustainable Development entitled "Megacities 2050: Environmental Consequences of Urbanization." The conference was organized in RUDN University, Moscow, Russia, on September 12–14, 2016. The conference aimed to provide a platform for discussion between different target groups involved in city sustainable development: environmental scientists, landscape architectures, urban planners, and policy-makers. We would like to thank more than 150 participants and 65 speakers who contributed with plenary, oral, and poster presentations. We wish to express our special gratitude to the authors who contributed to these proceedings. The proceedings include an introduction and 17 research papers, which were selected by the scientific committee with additional help of external expert reviewers from 31 submissions. The authors were asked to consider the reviewers comments and make all necessary edits to improve the quality of the papers.

The conference was organized under the umbrella of the "RUDN University program "5-100" and the "Erasmus+ Jean Monnet project "European traditions in governance, design and environmental management of megacities: search for solutions (EDEMS)." We would like to express our gratitude to RUDN University and Erasmus+ program for organizational and financial support. We are also deeply grateful to the many people who put essential efforts to ensure this successful conference: keynote speakers, members of organizing and scientific committees, sessions conveners and moderators, reviewers, and technical editors. We wish to express our sincere thanks to Dr. Michael Leuchner, Publishing Editor, earth sciences, geography and environment, and Rajan Muthu, Project Coordinator, for their help and cooperation.

We hope that these proceedings will serve as a valuable reference for researchers, practitioners, and policy-makers in the related fields.

Viacheslav I. Vasenev
Elvira Dovletyarova
Zhongqi Cheng
Riccardo Valentini

Organization

Committee

Chief Patron

Nur S. Kirabaev — Peoples' Friendship University of Russia (RUDN), Russia

Scientific Program Chair

Riccardo Valentini — Euro-Mediterranean Center on Climate Change, Italy

Organizing Co-chair

Viacheslav I. Vasenev — Peoples' Friendship University of Russia (RUDN), Russia

Scientific Program Committee

Elvira A. Dovletyarova — Peoples' Friendship University of Russia (RUDN), Russia

Vadim G. Pluyshchikov — Peoples' Friendship University of Russia (RUDN), Russia

Jetse Stoorvogel — Wageningen University

Zhongqi Cheng — City University of New York

Carlo Calfapietra — Institute of Agro-Environmental and Forest Biology (IBAF)

Ivan I. Vasenev	Peoples' Friendship University of Russia (RUDN), Russia
Tatiana A. Fedorova	Peoples' Friendship University of Russia (RUDN), Russia

Organizing Chair

Vadim G. Pluyshchikov	Peoples' Friendship University of Russia (RUDN), Russia

Organizing Co-chair

Elvira A. Dovletyarova	Peoples' Friendship University of Russia (RUDN), Russia

Organizing Committee

Viacheslav I. Vasenev	Peoples' Friendship University of Russia (RUDN), Russia
Carlo Calfapietra	Institute of Agro-Environmental and Forest Biology (IBAF)
Anna Paltseva	City University of New York
Elena Zaykova	Peoples' Friendship University of Russia (RUDN), Russia

Organizing Secretary

Tatiana A. Fedorova	Peoples' Friendship University of Russia (RUDN), Russia

Organized by

Contents

Signal and Image Processing

MEGACITIES 2050: From Urbanization Risks Towards Sustainable Urban Development

V.I. Vasenev[1,3(✉)], Z. Cheng[2], J.J. Stoorvogel[3], E.A. Dovletyarova[1], R.A. Hajiaghayeva[1], and V.G. Plyushchikov[1]

[1] Department of Landscape Design and Sustainable Ecosystems, Agrarian-Technological Institute, RUDN University, Moscow, Russia
vasenyov@mail.ru
[2] Department of Earth and Environmental Sciences, Brooklyn College of the City University of New York, Brooklyn, (NY), USA
[3] Soil Geography and Landscape Group, Wageningen University, Wageningen, The Netherlands

Abstract. Urbanization is a long-term global trend, responsible for substantial environmental changes. At the same time, urban ecosystems are vulnerable and their adaptation to the ever-changing environment is necessary to sustain essential functionality and important ecosystem services. Sustainable urban development demands the integration of innovative green technologies and nature-based solutions in urban management, which is only possible through a collaboration and participation of all stakeholders including scientists, landscape designers, civil engineers, policy makers, and all citizens.

Keywords: Urban ecosystems · Megapolis · Monitoring · Environmental management · Green infrastructure · Urban soils · Ecosystem services

Globally, urban areas grow rapidly with more than two thirds of the world population expected to live in cities by 2050 [1, 11]. Urbanization influences the environment and may contribute to e.g., climate change, soil degradation and biodiversity reduction. At the same time, urban ecosystems are very sensitive to global changes, and their adaptation is necessary to sustain essential functionality and important ecosystem services [5].

Historically, urbanization was mainly studied as a potential environmental threat, resulting in soil, water, atmospheric and forest degradation and biodiversity loss. The unfavorable ecological state of urban environments was documented by the beginning of the 21st century [3, 8]. An established urban ecosystem strongly differs from a natural or agricultural ecosystem. Urban ecosystems are characterized by the human modified and often artificial landscapes with considerable anthropogenic disturbances (e.g., environmental pollution, soil sealing, waste disposal). Cities generally consume much more energy than they generally provide, resulting in the emissions of heat, (airborne and waterborne) contaminants and greenhouse gases. With the continued increase of global urban population, novel concepts like 'sustainable cities' have emerged. The concept of urban sustainability resulted in the design of model or ideal cities, for example, 'emission free' cities [6] and 'climate adapted' cities [7] which view

V.I. Vasenev et al. (eds.), *Megacities 2050: Environmental Consequences of Urbanization*, Springer Geography, https://doi.org/10.1007/978-3-319-70557-6_1

urban areas as source of unique natural and urban-specific resources, rather than an environmental threat.

The international conference Megacities 2050 aimed to find solutions for environmental problems of modern megapolises and to maximize the capacity of urban ecosystems to support specific ('natural') functions and services. The conference proceedings introduce urban ecosystems, considering their spatial variability, temporal dynamics, environmental risks and potentials to provide important functions and ecosystem services. The volume includes 18 papers, describing different components of urban ecosystems (e.g., air, soil, vegetation and biota) and covering different aspects of environmental monitoring, assessment and management in megacities.

The general concept of megacities as diverse and complex ecosystems is presented in the first paper "Urbanization of Bioshphere: from Mega- to Ecopolises". The subsequent papers are organized into four different thematic sections: (i) air quality and greenhouse gases (GHGs) emission (papers 2 and 3), (ii) urban soils at multiple-scales (papers 4 to 9); (iii) urban forests and green infrastructure (papers 10 to 13); and (iv) advanced technologies in monitoring, modeling, designing and management of urban ecosystems (papers 14 to 18).

Maintaining air quality, carbon sequestration and mitigating global warming and climate changes by reduced GHGs emissions are key servives provided by urban ecosystems. The supply of these services for the cases of Naples (Italy) and Moscow (Russia) is discussed by the papers in Section 1. Urban soils are key for regulating healthy urban ecosystems. Ecosystem services and functions provided by urban soils impact the environment, and human health and wellbeing [4]. Urban soils that form conditions and features differ principally from natural and agricultural soils, but their functions and services remain poorly quantified [9]. Recently, there has been increased attention and interest in understanding the capacity of urban soils to support specific functions and services [5, 7, 10]. Currently, urban soils face a paradox where on one hand it is of the highest value for property development, and on the other hand being almost totally ignored with regard to the ecosystem servies they can provide [4]. Different aspects of monitoring and assessment of urban soils at multiple scales from local and city level (Rostov in Russia and New York in USA) to regional and global scales are discussed in Section 2. Similar problems (e.g., contamination with heavy metals) were presented for urban soils located at different climates and vegetation zones (e.g., Yamal in arctics, Bashkortostan in steppes and New York in the humid continental/temperate), providing a unique opportunity for comparative assessments. Section 3 focuses on green infrastructure as the main tool to integrate nature-based solutions into urban design and management. Finally, Section 4 promotes a range of technologies to monitor and manage urban ecosystems, including biotesting, decision-support systems and ecological engineering.

The conference received feedback from a broad and multi-disciplinary audience, including the scientific community, municipal services, the environmental protection agency and other stakeholders working in urban management and greenery. Such a multi-disciplinary discussion is an essential step towards sustainable urban development, because implementation of innovative technologies and nature-based solutions relies on a collaboration of all interested stakeholders for the purpose of smart urban management.

Acknowledgments. The RFBR project NK 15-34-70003 and Jean Monnet Project EDEMS supported the conference and the research.

References

1. FAO: Climate-Smart Agriculture. Sourcebook, E- (2013). ISBN 978-92-5-107721-4
2. Gómez-Baggethun, E., Barton, D.N.: Classifying and valuing ecosystem services for urban planning. Ecol. Econ. **86**, 235–245 (2013)
3. McKinney, M.L.: Urbanization as a major cause of biotic homogenization. Bio. Conserv. **127**, 247–260 (2006)
4. Morel, J.L., Chenu, C., Lorenz, K.: Ecosystem services provided by soils of urban, industrial, traffic, mining, and military areas (SUITMAs). J. Soil Sed. **15**, 1659–1666 (2015)
5. Pickett, S.T.A., Cadenasso, M.L., Grove, J.M., Boone, C.G., Groffman, P.M., Irwin, E., Kaushal, S.S., Marshall, V., McGrath, B.P., Nilon, C.H., Pouyat, R.V., Szlavecz, K., Troy, A., Warren, P.: Urban ecological systems: scientific foundations and a decade of progress. J. Environ. Manag. **92**, 331–362 (2011)
6. Pickett, S.T.A., Cadenasso, M.L., Grove, J.M., Groffman, P.M., Band, L.E., Boone, C.G., Burch, W.R., Grimmond, C.S.B., Hom, J., Jenkins, J.C., Law, N.L., Nilon, C.H., Pouyat, R. V., Szlavecz, K., Warren, P.S., Wilson, M.A.: Beyond urban legends: an emerging framework of urban ecology, as illustrated by the Baltimore Ecosystem Study. Bioscience **58**, 139–150 (2008)
7. Raciti, S.M., Groffman, P.M., Jenkins, J.C., Pouyat, R.V., Fahey, T.J., Pickett, S.T.A., Cadenasso, M.L.: Accumulation of carbon and nitrogen in residential soils with different land-use histories. Ecosystems **14**, 287–297 (2011)
8. Stroganova, M.N., Myagkova, A.D., Prokofieva, T.V.: The role of soils in urban ecosystems. Eurasian Soil Sci. **30**, 82–86 (1997)
9. Vasenev, V.I., Ananyeva, N.D., Makarov, O.A.: Specific features of the ecological functioning of urban soils in Moscow and Moscow oblast. Eurasian Soil Sci. **45**, 194–205 (2012)
10. Vasenev, V.I., Stoorvogel, J.J., Vasenev, I.I.: Urban soil organic C and its spatial heterogeneity in comparison with natural and agricultural areas in the Moscow region. CATENA **107**, 96–102 (2013)
11. United Nations.: World Urbanization prospects: the 2007 revision. United Nations Department of Economic and Social Affairs, Population Division. New York, USA (2008). http://www.un.org/esa/population/unpop.htm

Urbanization of Biosphere: From Mega- to Ecopolises

Dmitry N. Kavtaradze(✉)

Biological Faculty, Moscow State University, Moscow, Russia
kavtaradze@mail.bio.msu.ru

Abstract. Sustainable development requires principle changes in goals, indicators and values urban areas to well-known principles from the fundamental work "Limits to Growth" [20]. Multidisciplinary and cross-sectorial teams of experts, planners and authorities need to elaborate jointly national and inter-regional strategies to the ecological mode of urbanization to turn dynamics of megapolises growth to ecopolises. New type of urban and ecopolis strategies as the core of urban strategy tested on the example of Korolev science city in Moscow Region by ecosystem restoration, evaluation of risks and calculation of resources required for implementation ecological master plan. This task force elaborated and designed with strong participation with the city council of Korolev by a joint Russian and Italian team of students and professors [15].

Keywords: Urbanization · Megapolis · Carrying capacity · Sustainable development · Ecopolis · Coherent development · Post-industrial · Urban risk

1 Industrial Approach to a Growth

Although researchers attempted to find solutions to the problems facing humanity during the last decades of the twentieth century, handling these problems continues to be an issue as the conclusions they drew questioned in the twenty-first century [1, 14, 20]. As no new common ground (paradigm) has emerged, Aurelio Peccei - the founder of the Club of Rome and Institute of System Researches - has said that one needs compelling arguments to "forecast a potential catastrophe in the coming decades" and that "there is a necessity in the Great change of the direction of human activities" [19]. Predictions were not accepted. Modem science cannot forecast evolutional changes in biosphere, which will ensure safety of human population [2, 20].

We need to open social, scientific and professional meaning of "sustainable cities." The concept of "Ecopolis" as urbanization strategy and experiment started in 1980's in USSR - developing human settlements that are coherent with regional ecosystems carrying capacity. Science starts with the determination of goals and terms. Science attempts to monitor and model dynamics, while philosophers and other thinkers try to understand trends to define limits of urbanization. After the Da Vinci Urban Project, several utopian projects have mostly been successful and generalized plans been developed. Finally, Garden Cities open new area for humans - eco-return to reasonable natural milieu. Industrialization since the 19th century has led to population super density urban sprawl, swallowing forests and agricultural areas in the countryside. This

V.I. Vasenev et al. (eds.), *Megacities 2050: Environmental Consequences of Urbanization*, Springer Geography, https://doi.org/10.1007/978-3-319-70557-6_2

has brought impacts on biosphere that not been predicted. A new profession, "eco-urbanist", has emerged through research and project practice design. During the past decades in Russia, we in fact eliminated long-term planning and strategy in human settlement policy.

2 Post-industrial Shifting from Growth to Development

The direction of urbanization in general has not been clear. Today we need to revisit the long-term values and support state strategy in urbanization design. Models should followed by the development of low risk scenarios in regional ecosystems capacity and local ethno-cultural context. In this way, important projects appeared: bioregionalism, eco-cities and ecolopolises as self-supporting settlements. In Russia several large-scale projects were fulfilled starting from the beginning of XX centuries in the vicinity of Moscow (beginning of XX c. on railway St. Piozorovskay) and in 1970–1990 in Science City's Pouschino, later in Korolev (previous Kaliningrad Moscow region).

Ecopolis approach tested in Old Kosino and Vologda. Megacities as another kind of human settlement are suggesting certain social metabolism where matter and energy flows. Human ecology issue is not on a short list of modem progress. Due to given scenarios of population growth [17], new limits of centralized management can be expected - systems will develop to super complex level with risks of unpredicted disasters - "normal accidents" [20]. Strategy of global urbanization is slowly moving toward a new integrated policy that is influenced by climate change, shifting to alternative energy, sea level rising, the decline of per capita plough land on the planet limited fresh drinking water supply for global citizen, urban stress, etc. [9]. Those different risks required the integration a system with the lowest integrative risks by means of new mode of urbanization - ecopolis. Our determination is – Ecopolis is a multifunctional unit in planning and it even partly produces vital resources from within, e.g., drinking water, food (fish, game products, cattle, wheat vegetables, and fruits), natural berries, mushrooms, and additional planting local plant species. City ecological service fulfilled through environmental and settlement monitoring, ecosystems restoration, ecological network connection among regional and Ecopolis ecosystems. Ecopolis as a project have started at the very beginning of 1980's in Science City Poushino in the Soviet Union era. Poushino Science-City was in some way the "town of future' as there was no industrial impact, only recreational. That impact was carefully studied in all seasons and it showed that recreational needs (e.g., walking, gathering flowers, fishing, hunting, mushrooms picking) actually use territory more than 600 times bigger than its own city territory.

Training and education of local deputies can improve official's vision of planning and possibly expand it to "geological time scale." Only after participation in long term modeling experiments and projects participants (volunteers, deputies, and local scientists) were motivated and ready to think and discuss future changes of landscape, city and human habits, and to model "common vision of desired future." Values of sustainable development has to start in human mind first [11, 14, 16, 22].

Local city natural reserves planned and workforce was dedicated officially to take care of them, including volunteers mostly with high education background.

The Ecopolis experience established a link between academic philosophy and everyday human practice. It finally realized the first steps of coherent development of regional ecosystems and urban milieu. This is the new meaning of modern cities: biodiversity and ecosystems restoration become the center of urban system dynamics modeling [15, 18, 21] on the way of sustainable satisfaction of human needs.

3 Multidisciplinary Approach to Avoid Misunderstanding

The term "Ecopolis" invented simultaneously with "Coherent Development of the Nature and the Society" [2] concept elaborated in the 1980s. Important to give description of the term "Ecopolis" – is a human settlement and its immediate surroundings metabolically coherent with natural regional ecosystems and offer fruitful social and cultural milieu. This theoretical ideal directs efforts of scientists, planners, citizen, politicians, health and ecological services to organize field polygons as application of the theory. "Fragments of the Ecopolis concept are recognizable in many cultural landscapes and cities as green areas, urban forests, streams and channels. Ecopolis is a human settlement of the future, whose ecological parameters are controlled and whose inhabitants culturally prepared for constant changes, both in the mode of life style and in nature" [2]. Research and experiment on mutual adaptation of man and nature started in Poushino (Moscow region) in 1980 and run for several decades, with extensive participation of local scientists, citizens and the administration of the Science City Project. Observing eco-cities in different countries practice, difficult to recognize scientific approach except for taking care of elements of city metabolism such as water and air quality, and urban infrastructure planning to save energy. Large-scale research results since 1980s was widely discussed [3, 11, 12]. As far as we know, the "Ecopolis concept" has not implemented in practice internationally, taking into account presentations at international conferences and two World Exhibitions (1985, 1987). Significant that independently, ecological research and open space design of human settlements on planets has been done in the by I. Gitelson in Krasnoyarsk laboratory in Russia. Worldwide known the unique experiment "Biosphere II" in Arizona supported by Eco-Technics Institute in London and is in operation until today. The eco-villages are small communes and popular in many countries and are managed without science background, just by using organic farming and follow agriculture techniques [8].

In Russia, science traditions often based on philosophy and biosphere theory invented by Vladimir I. Vernadsky. His theory was one of the cornerstones to urban areas review and improvement. The term "ecopolis" [2] covers the historical experience of urban design policies and instill in the practice the modern understanding of humankind vital dependence on the biosphere conditions and technical irreplaceable ecosystems service.

Later the catch-all title "sustainability" has become popular lexicon of environmental projects. As it was analyzed in the School of Architecture at Alghero (Sardinia - Italy), projects often referred to as descents of scale, particular in the architectural dimension and construction, or in qualitative and quantitative methods [3]. Discussion around terms and meanings we value as normal processes in the sense given be

Herman Daly to "Strong" sustainability opposite to the "non-substitutability of natural capital by "artificial" capital (human, social and economic). The sum of natural and anthropogenic capital in other words be kept at a constant value, or each component can be kept constant. It is reasonable, if we think that the two types of capital are interchangeable with each other, need to be proved" [4–6]. The latter is could be accepted if we think that the natural and manufactured capital is complementary, what is not obvious.

Therefore, those capitals should be operate coherently because the productivity of one depends on the other [2, 12].

Land use and development are emergent properties of a complex system – they are not leverage points in themselves. The first step in research and education efforts done jointly with local institutions, reasonably elaborating observation of domestic design and construction traditions. Important to figure out their "environmental reasons" including their bioclimatic function role and mediating energy flows).

The goal is to identify modes of development and regeneration of dwelling that fit these approaches. This exploratory work also accompanied by research and educational activities in close connection with energy supply by renewable sources. Important to recovery traditions of local construction with low environmental impact, including practice of "eco-friendly and durable building materials with significantly longer lifecycle" [7].

4 SD Principles in Local Planning

Only very recently environmental risks start to be discussed and taken into account referring to ecosystems carrying capacity and their general conditions. Large-scale experiments as well as practically tested guide for planning needed with continuous long term monitoring are crucial to value theory "Coherent Development of the nature and the Society" [2]. At the same time, basic principles of Sustainable Development (SD) need implementation:

1. Climate change as factor influencing mitigating potential impacts in future by modeling urban planning and smart houses using;
2. Addressing more projects to aquatic, marine systems, and underwater settlements;
3. Taking into consideration industrial load, field ecology data, biodiversity and general environmental issues;
4. Using information technology, system dynamics to design simulation models highly urbanized territories metabolism, and management scenario.

Each principle represent a part of simultaneous actions in the system eco-urban planning finally focused on a certain objects (Table 1).

From these principles, several core objectives and professional fields of activity follow:

Monitor, protect and restore highly urbanized systems. That suggest rehabilitation of the local reserves, parks, water bodies improving environmental structure of the territory, carrying capacity, and efforts to maintain them. Finally improving connectivity and avoiding fragmentation.

Table 1. Management and coherent action principles for ecological Korolev sustainable development

Six principles of coherent SD	Eco-logical actions on all levels planning
Protect, restore and enhance existing ecosystems and network of green areas, water bodies	Ecological infrastructure maintain at all levels. Policy of coherent of landscape mitigation and life style Fragmentation and improve ecological network Mitigation fragmentation of different landscape levels
Closing the matter and energy cycles: environmental resources management	In climate change dynamic prevent flooding Improve carrying capacity recreational places. Suggest planning new one Suggest management to complete waste and water cycles
Natural resources management	Maintain load and monitoring the reserves as the system network Establish green service Provide visit centers, materials for education & recreation Organize the collaboration and activity of professional corps and volunteers on levels of activity
Boundary of urban- countryside planning and design	Support and fulfil transforming edge zones of urban ecosystems to countryside ecosystems as joint Econet practicing as an milieu connecting urban-countryside and regional ecosystems Restore, local and regional land fields, abandoned, residual lands for new functions Support practice of urban agriculture
Sustainable housing	Save and use local renewable resources with smart buildings Reuse and return to social use abandoned buildings balance density of citizen population and social needs
Risk management	Invent integrated risk management of land-use: hydrological, geological, territory planning, industrial, postindustrial activity, civil activity Invent monitoring of natural and urban milieu risks dynamics, established public source of on-line information, including health risks, quality of air and water and other vital resources. Invent maps of urban risks interrelations from geological to health and criminal and integrated risk, imparting natural disaster

Risk management deals with either the adaptation to storm water and flooding, or the industrial risk nano-particle air pollution, as space technologies are located in Korolev.

Fulfilling the EKorolev project as social practice used different strategy and tactics mainly based on human resources.

Implemented by the City Green Service, the most important was to improve elements and certain places that are public, open and coherent with the goal and values of the project. In addition, highways improved and reduced the traffic load from national park. Even good projects are not easy to implement in city. As we know municipal government often practice independent tactics and more important tasks that needed to complete urgently. Finally, measures addressed to sustainable project development loose importance.

The Master Plan was assembled together with other sketches, graphics, photo-simulations, and suggestions sheets, in which every single idea is represented.

PowerPoint presentation also provided for the public – that make it easier to understand and discuss project with images, slides and animations.

5 Conclusion

The final report was presented as hierarchy of concept, strategy and project plan to colleagues, from universities, experts and finally to administration and mayor of Science City Korolev. Latter the Russian-Italian team of the EKorolev project was designing Master Plan of Geological Park on Sardinia, materials published bilingual in Moscow [3, 12]. Project is an example of principles that request simultaneous actions, controversially to "step by step" practice that bring postponed conflicts and controversial effects [1, 15]. The meaning of the EKorolev project is multicultural, multidisciplinary as City is evaluate as the most complex system created by humans. Such complex systems need model design to choose the best scenario with the lowest risk. "Build inclusive, safe and sustainable cities and human settlements" [8, 10, 13], cities need to take care of local ecosystems as it was highlighted in the United Nations General Assembly in 2015. The need in long-term Urbanization Strategy of Russia for new and old settlements, importance "ecopolis" approach discussed on meetings Public Chamber of the Russia in 2014–2017 years.

Acknowledgments. Cross-cultural and multidisciplinary research and art of planning was possible due to creative students of Russian and Italian Universities and EKorolev master plan designed by talented Alessandra Casu, MSc, PhD., University of Alghero, Italy and consultations with world known methodologist and practical planner of sustainable city development Gwendolyn Hallsmith from Vermont, USA.

References

1. Beck, U.: Risk Society: Toward a New Modernity. Sage, London (1992)
2. Brudny, A., Kavtaradze, D.: Ecopolis. Principles and Problems. ONTI, Pouschino (1981). (in Russian). Brudny, A., Kavtaradze, D.: The coherent development of the nature and the society. The International Exposition. Vneshtorgisdat, Tsukuba, Moscow, Japan (1985). Brudny, A.A., Kavtaradze, D.N.: "Ecopolis": prognosis turned into project. Social Science, no. 2 (1984)
3. Casu, A.: Approaches to sustainable planning. In: Kavtaradze, D.N., Casu, A. (eds.) EKorolev. Projects of Regional Sustainable Development. The Report on Russian-Italian and Italian-Russian Student Workshops in 2007–2008, Educational monograph, UN Decade, Education for Sustainable Development, Acropol, Moscow (2011)

4. Daly, H.: Beyond Growth. The Economics of Sustainable Development. Beacon Press, Boston (1996)
5. Daly, H.E.: Toward some operational principles of sustainable development. Ecol. Econ. **2**, 1–6 (1990)
6. Daly, H.E.: Operationalizing sustainable development by investing. In: Natural Capital, ISEE Conference, Stockholm (1992)
7. Deelstra, T.: Public participation in improving urban life and environment. In: Cities and Ecology, MaB Program, Suzdal, USSR, Collected Reports, vol. 2, pp. 153–167 (1988)
8. Gerzberg, L.: Ecosettlements in Russia. In: International Conference. School of Public Administration, Moscow State University, Public Administration, Russian Federation in Modern World. Oral presentation (2015, in print). Ignatieva, M.E.: Ecopolis-search for sustainable cities in Russia. In: The Sustainable City II: Urban Regeneration and Sustainability, pp. 53–61. WIT Press (2002)
9. Indicators: European Environment Indicators (2016). http://www.eea.europa.eu/data-and-maps/indicators/#parent-fieldname-title
10. ISS (2014). http://www.un.org/sustainabledevelopment/cities/
11. Kavtaradze, D.: Ecopolis and new urban functions. In: Conference on Proceedings, IALE Congress, China (2011)
12. Kavtaradze, D., Casu, A.: "EKorolev"- Dimensions in Urban and Countryside Planning Sustainable Development/NISPACE HSE, November 2015
13. Kavtaradze, D.N.: Public and professional education in urban ecology, perception and evaluation of urban environment quality. In: Bonnes, M. (ed.) Proceedings of the International Symposium on a Pluridisciplinary Approach in the European Context Rome, Project 11. MAB, Italia (1991)
14. Kavtaradze, D.N., Boukvareva, E.N., Sidorenko, V.N.: Using of system dynamics and geographical information systems for ecological nets planning. In: Proceedings of System Dynamic Society Conference, Norway, July 2000, p. 213 (2001)
15. Kavtaradze, D.: Urbanizatsij of the Biosphere/Problemi okrushauschei sredi i prirodnih resursov. Obsornaj informatsij, no. 7. VINITI, Moscow (2005)
16. Kavtaradze, D., Boukvareva, E., Sidorenko, V.: 'Econet-ABC': Imitatsionnaya upravlencheskaya igra po sozdaniyu seti osobo ohranyaemyh prirodnyh territoriy regiona ['Econet-ABC': Simulation game for Creation of Regional Natural Area of Preferential Protection]. Izd-vo Che-Ro, Moscow (2005)
17. Meadows, D.H., Meadows, D.L., Randers, J., Behrens III, W.W.: The Limits to Growth. Universe Books, NYC (1972)
18. Numata, M.: Water-oriented approach to urban ecosystems. In: Cities and Ecology, MaB Program, Suzdal, USSR, Collected Reports, vol. 2, pp. 39–46 (1988)
19. Peccei, A.: The Chasm Ahead. T MacMillan Company Collier-MacMillan ltd., London (1969)
20. Perrow, C.: Normal Accidents: Living With High Risk Technologies, Revised edn. Princeton University Press, Princeton (2009)
21. Sukopp, H., Henke, H.: Nature in towns: a dimension necessary for urban planning today. In: Cities and Ecology, MaB Program, Suzdal, USSR, Collected Reports, vol. 2, pp. 30–36 (1988)
22. Turner, G.: Is Global Collapse Imminent? Research papers Series N 4. Melburn Sustainable Society Institute. University of Melburne, August 2014

Measuring the Role of Urban Vegetation on Air Quality: The Case Study of the Real Bosco di Capodimonte in Naples, Italy

Raffaela Esposito[1,2], Gabriele Guidolotti[1], Emanuele Pallozzi[1], Corrado Leone[1], Michele Mattioni[1], Giovanni De Simoni[1], and Carlo Calfapietra[1,3(✉)]

[1] Institute of Agro-Environmental and Forest Biology (IBAF), National Research Council of Italy (CNR), Porano, Naples, (TR), Italy
carlo.calfapietra@ibaf.cnr.it
[2] Department for Innovation in Biological, Agro-Food and Forest Systems, University of Tuscia, Via San Camillo de Lellis, Viterbo, Italy
[3] Global Change Research Institute, Academy of Sciences of the Czech Republic, Brno, Czech Republic

Abstract. Urban vegetation, particularly urban forests are a crucial component of green infrastructure in our cities. A strong interest is related to the ecosystem services provided to both people and the environment including CO_2 sequestration and air quality improvement. However few *in situ* experimental studies exist on these topics while most of the estimations derive from modelling exercise. Preliminary results of an innovative eddy covariance site located in the Real Bosco di Capodimonte within the urban area of Naples (Italy) are presented here.

Keywords: Urban forest · Green infrastructure · Air quality · Trees · Eddy covariance · Air pollution

1 Introduction

Anthropogenic activities have strongly altered the global environment, with effect at both global and local scale. At global scale, one of the most investigated aspects is the atmospheric concentration of carbon (C), in particular carbon dioxide (CO_2) increase - and the resulting climatic global changes [1]. Cities cover a minimal part of the world's dry land surface [2], but release more than 70% of the total emissions of anthropogenic CO_2 [3]. At local scale, the human impacts can be even harsher: according to the recent report of the World Health Organization, globally up to the 80% of people living the in urban areas are exposed to pollutants concentration levels that undermine the health of the citizen [4]. Vegetation, especially forests, has a prominent role on the regulation and mitigation of these issues. They fix CO_2 through the photosynthetic process, while a part of it returns to the atmosphere by means of respiratory autotrophic and heterotrophic processes, and a part is stored for long periods as living biomass and in the soil. Moreover, vegetation can have a direct absorption of air pollutants by leaf surface [5].

V.I. Vasenev et al. (eds.), *Megacities 2050: Environmental Consequences of Urbanization*, Springer Geography, https://doi.org/10.1007/978-3-319-70557-6_3

On this basis, a greater attention should be focused on urban forests, that for instance in Italy account for about 43000 ha, with a mean area of 2.2 ha each [6]. A better knowledge of the role of urban forests on the interaction with CO_2 and the other air pollutants (e.g. O_3) could significantly help the local authorities in taking decisions to maximize their services and to understand how many benefits they might provide [7]. Today, this is of central interest, also in relation to the Covenant of Majors signed in 2008, an agreement that obliges European cities to establish an Action Plan to reduce their CO_2 emissions by over 20% through activities and practices including trees' plantation in urban areas [8].

Often, for the heterogeneity of urban landscape and the elevated number of factors involved, CO_2 and air pollutants removal is often estimated by models [9]. In any case, the most accurate way to assess the role of urban forest on air quality is to measure it directly. Among the various methodologies, Eddy Covariance (EC) technique is the most appropriate to allow instantaneous, continuous, precise, non-invasive and wide-scale measurement study of fluxes exchanged between biosphere and atmosphere. Here, we present a unique worldwide facilities and some preliminary results based on the application of EC in an urban forest, where fluxes of CO_2 and H_2O are simultaneously measured along with ozone (O_3), which is a harmful pollutant for plants and humans. The station was established in the "*Real Bosco di Capodimonte*", a large urban park within the large city of Naples in Italy [10].

2 Materials and Methods

2.1 Site Description

Naples is one of the most densely populated Italian cities, with about one million inhabitants [11]. Counting day-trippers and commuters, it reaches four million citizens every day.

The Real Bosco of Capodimonte benefits from a superb position: located on top of a breezy hill that overlooks the entire gulf of Naples; it is visible from most of the city. Because of these features, in 1734 Charles of Bourbon chose Capodimonte as his game reserve. The park is characterized from sixteen structures, including the church of San Gennaro, above which the flux tower is located (coordinates: lat. 40.8741, long. 14.2504, 133).

The entire area of the Real Bosco is characterized by Mediterranean climate [12], and covers about 134 hectares. Tall trees (more than 150,000), mainly *Quercus ilex* and *Pinus pinea* are alternated with exotic species, shrubby and large meadow areas.

2.2 Eddy Covariance System

As described by Guidolotti et al. [10], a meteorological station to measure global radiation, precipitation, air pressure, relative humidity and temperature was installed at the end of 2014. Moreover, in the same place an EC tower was established composed by a 3-D sonic anemometer (WindmasterPro, Gill, United Kingdom), a CO_2/H_2O closed path infrared gas analyzer (LI-7200, LI-COR, Lincoln, NE, USA), a Fast Ozone

Analyzer (FOA, Sexstant technology, NZ), and a Dasibi 1008 HC U.V. photometric ozone analyzer (Dasibi environmental, CA). All the data were collected at 10 Hz by means of two Campbell (CR1000 and CR6) and LI-7550 dataloggers. Fluxes of CO_2, H_2O, O_3, latent heat (LE), sensible heat (H), and momentum were processed using the EddyPro (EP) software [13].

3 Results

Here we show some preliminary results based on 121 days at beginning of 2015 to understand the area covered by the EC tower measurements (Fig. 1) and a 3 days frame of CO_2 and O_3 fluxes (Fig. 2). The prevailing wind directions were south-west and north-west, which accounted together for more than 50% of the total (Fig. 1a). The wind influenced the distribution of the exchange source area (footprint). Indeed the analyzed footprint (upwind area that generate the flux) is mainly characterized by a mixed Mediterranean forest dominated by *Quercus ilex* with a mean height of 22 m, and meadows are present, mainly composed by species of *Trifolium* and *Medicago*. At the beginning of 2015 more than 50% of the data derived from the meadows areas while the remaining from the holm oak forest and from the residential area, in particular when the source area went beyond the 150 m in the west direction. In addition, no marked differences were observed in the footprint during day and night time (Fig. 1b, c), probably for the elevated wind turbulence during the night due to the seaside location of the site characterized by breezes. Guidolotti et al. [10] suggested that long term measurements will provide enough data to analyze it following a "split-footprint" approach.

In Fig. 2, CO_2 and H_2O fluxes (panel a) and O_3 fluxes (panel b) are shown. Although the period presented is very short, the shown data provides relevant information about the function of the urban park of the city of Naples underlining the impact of the vegetation on the urban atmospheric composition. The CO_2 fluxes measured in the experimental site are similar to what is reported for other natural [15] and

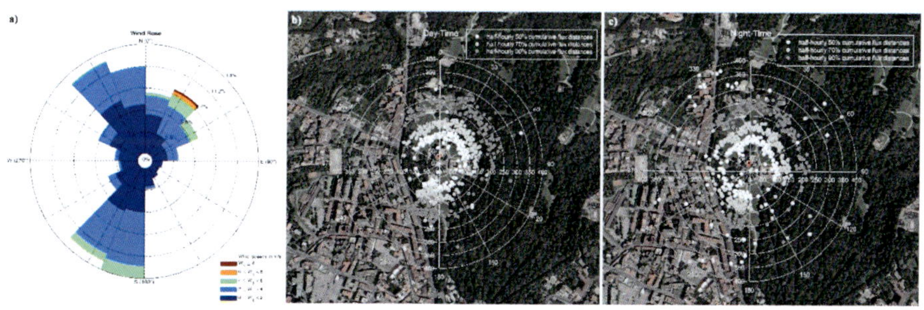

Fig. 1. Wind distribution for 121 days from the beginning of year 2015 (panel a) at the EC station of the Real Bosco di Capodimonte in Naples. Half-hourly cumulative fluxes representing the 50% (with circles), 70% (light grey circles) and 90% (dark grey circles) during day (panel b) and night (panel c) at the end of March. Cumulative fluxes distances were calculated using the footprint model of [14].

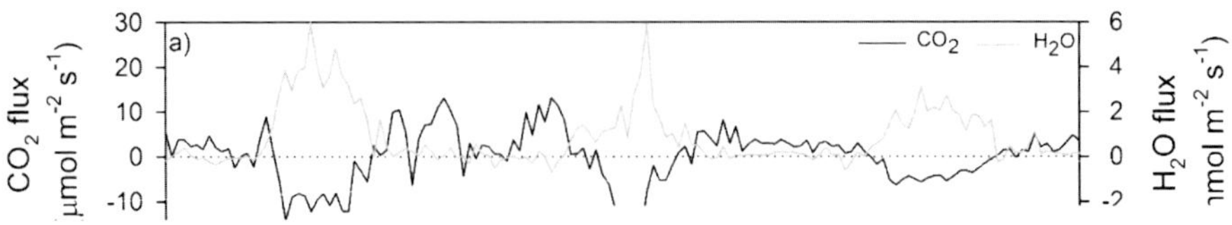

Fig. 2. Temporal course of half-hourly fluxes of CO_2 and H_2O (a), O_3 (b) for the end of March 2015 at the EC station of the Real Bosco di Capodimonte in Naples. The dotted line represents constantly the zero, in order to evidence the absorption (0 < values) and the emission (0 > values) according to the micrometeorological convention.

peri-urban forests in Mediterranean area [16]. Moreover, the opposite daily trend between CO_2 and H_2O is proving the effect of vegetation carbon absorption and evapotranspiration during the day (Fig. 2a), suggesting the potential carbon-sink of Capodimonte as found in other urban parks [17]. The O_3 fluxes, mostly negative during the day (Fig. 2 b), when evapotranspiration (and thus stomata) is higher are proving the effect of the urban park on O_3 deposition. Despite the short time frame presented here, the recorded O_3 fluxes are in the range reported for a *Quercus ilex* forest of central Italy [16], where the authors evidenced the role of the urban environment in depleting the O_3 concentration, due to the O_3 reaction with other anthropogenic trace gases [10, 16].

Further analysis with longer dataset will be carried out at the site to better understand the role of urban park on both carbon absorption and O_3 deposition, accounting also for park structure (trees vs meadows). All this information will give the opportunity to better understand the role of the urban park and its structure on urban air quality at both global and local scale.

4 Conclusion

Our results suggest the suitability of EC in urban forests. Moreover, we promote the idea of an urban forest EC research network, such as Fluxnet and ICOS networks, given the target to realize inventories of urban forests for carbon stocks estimations. Moreover, these kinds of stations will allow to increase our understanding on the relationships between anthropogenic and biogenic sources and their implications for air quality in cities. A network of these stations would be highly representative from the scientific point of view and would provide crucial information for the planning and management of green areas and urban forests in our cities.

References

1. IPCC: Climate change 2014: impacts, adaptation, and vulnerability. In: Part A: Global and Sectoral Aspects. Contribution of Working Group II to the 5th Assessment Report of the Intergovernmental Panel on Climate Change. Cambridge University Press, Cambridge (2014)
2. Satterthwaite, D.: Cities' contribution to global warming: notes on the allocation of greenhouse gas emissions. Environ. Urban. **20**, 539–549 (2008)
3. Canadell, J., Ciais, P., Dhakal, S., Le Quéré, C., Patwardhan, A., Raupach, M.: The human perturbation of the carbon cycle. In: Persic, A. (ed.) The Global Carbon Cycle, UNESCO-SCOPE-UNEP Policy Briefs Series. Paris (2009)
4. World Health Organization. http://www.who.int/mediacentre/news/releases/2016/air-pollution-estimates/en
5. Nowak, D., Crane, D., Stevens, J.: Air pollution removal by urban trees and shrubs in the United States. Urb. For. Urb. Green. **4**, 115–123 (2006)
6. Corona, P., Agrimi, M., Baffetta, F., Barbati, A., Chiriaco, M., Fattorini, L., Pompei, E., Mattioli, W.: Extending largescale forest inventories to assess urban forests. Environ. Monit. Assess. **184**, 1409–1422 (2011)
7. Livesley, S., McPherson, E., Calfapietra, C.: The urban forest and ecosystem services: impacts on urban water, heat, and pollution cycles at the tree, street, and city (2015)
8. Archer, D., Dodman, D.: Making capacity building critical: Power and justice in building urban climate resilience in Indonesia and Thailand. Urban. Clim. **14**, 68–78 (2015)
9. Guidolotti, G., Salviato, S., Calfapietra, C.: Comparing estimates of EMEP MSC-W and UFORE models in air pollutant reduction by urban trees. Environ. Sci. Poll. Res. **23**, 19541–19550 (2016)
10. Guidolotti, G., Calfapietra, C., Pallozzi, E., De Simoni, G., Esposito, E., Mattioni, M., Nicolini, G., Matteucci, G., Brugnoli, E.: Promoting the potential of flux-measuring stations in urban parks: an innovative case study in Naples. Italy. Agr. For. Meteorol. **233**, 153–162 (2017)
11. ISTAT. http://www.istat.it/storage/urbes2015/napoli.pdf
12. La Valva, V., De Natale, A.: Prime osservazioni sulla flora urbana di Napoli. Allionia. **32**, 215–217 (1994)
13. Fratini, G., Ibrom, A., Arriga, N., Burba, G., Papale, D.: Relative humidity effects on water vapour fluxes measured with closed-path eddy-covariance systems with short sampling lines. Agric. For. Meteorol. **165**, 53–63 (2012)
14. Kljun, N., Calanca, P., Rotach, M., Schmid, H.: A simple parameterisation for flux footprint predictions. Bound. Layer Meteorol. **112**, 503–523 (2004)
15. Reichstein, M., Tenhunen, J., Roupsard, O., Ourcival, J., Rambal, S., Dore, S., Valentini, R.: Ecosystem respiration in two Mediterranean evergreen holm oak forests: drought effects and decomposition dynamics. Funct. Ecol. **16**, 27–39 (2002)
16. Fares, S., Savi, F., Muller, J., Matteucci, G., Paoletti, E.: Simultaneous measurements of above and below canopy ozone fluxes help partitioning ozone deposition between its various sinks in a Mediterranean Oak Forest. Agric. For. Meteorol. **198**, 181–191 (2014)
17. Kordowski, K., Kuttler, W.: Carbon dioxide fluxes over an urban park area. Atmos. Environ. **44**, 2722–2730 (2010)

Urban Soil Respiration and Its Autotrophic and Heterotrophic Components Compared to Adjacent Forest and Cropland Within the Moscow Megapolis

V.I. Vasenev[1,2(✉)], S. Castaldi[3], M.M. Vizirskaya[1], N.D. Ananyeva[4], A.S. Shchepeleva[1,2], I.M. Mazirov[1], K.V. Ivashchenko[2,4], R. Valentini[5], and I.I. Vasenev[1,2]

[1] Laboratory of Agroecological Monitoring, Ecosystem Modeling and Prediction, Russian State Agricultural University, Timiryazevskaya str., 49, Moscow 127550, Russia
vasenyov@mail.ru, Vasenev_vi@pfur.ru

[2] RUDN University, Miklukho-Maklaya str., 6, Moscow 117198, Russia

[3] Dipartimento di Scienze e Tecnologie Ambientali, Biologiche e Farmaceutiche, Seconda Università di Napoli, via Vivaldi 43, 81100 Caserta, Italy

[4] Institute of Physico-chemical and Biological Problems in Soil Science, RAS, Institutskaya str., 2, Pushchino, Moscow Region 142290, Russia

[5] Tuscia University, Via del Paradise 47, Viterbo, Italy

Abstract. Urbanization is a key land-use change pathway, increasing urban population and resulting in abandonment of rural areas. Urbanization alters profoundly soil properties and functions, including soil respiration [90]. Soil respiration includes heterotrophic (microbial) and autotrophic (root) components. Both are driven by biotic and abiotic factors. Soil respiration and its components in urban ecosystems remain poorly known. In the present study, the spatial and temporal variability of total soil respiration (Rs) and its components were analyzed for different ecosystems included in the Moscow megalopolis area. In particular, highly impacted areas of urban green lawns were compared to arable lands and urban forest sites. Respiration fluxes were monitored during the whole vegetation period. An average Rs was significantly higher for the most disturbed sites, compared to more natural sites. For all the sites, Rh was the dominant component of soil respiration. We report the highest heterotrophic respiration ratio to microbial C (qCO_2 = Rh/C_{mic}) for the lawn land use, followed by arable sites and forest sites, characterized by the lowest qCO_2. An average Ra contributed to total Rs only to a minor extent (26%) and increased in all study sites along the season. Ra absolute values and contribution to Rs were similar for different land use types.

Keywords: Root respiration · Microbial respiration · Urban lawns · Urban forest · Microbial biomass

V.I. Vasenev et al. (eds.), *Megacities 2050: Environmental Consequences of Urbanization*, Springer Geography, https://doi.org/10.1007/978-3-319-70557-6_4

1 Introduction

Soil is the largest terrestrial pool of actively cycling C (C), containing about 3000 Pg (10^{15} g) C [38, 42, 82]. Although soils are the major C sinks, they also have potential to emit C through soil respiration [14, 30, 92], which represents the predominant terrestrial CO_2 efflux from land to atmosphere [70, 76]. Carbon dioxide (CO_2) emissions due to soil respiration vary between ecosystem type and among biomes [16, 20, 27, 36], and are strongly influenced by land-use change [26, 34, 35]. Significant increases of CO_2 emissions have been reported after deforestation [31, 84], and during the first years after conversion of natural areas into croplands [21]. Drastic changes also occur when abandoned agricultural lands are recolonized by natural vegetation [48, 92].

Much less is known about urban expansion influence on soil C fluxes. Urban areas are highly heterogeneous. The percentage of the impervious areas varies in different parts of the city and functional zones and may differ from 7 to 90% with an average of 50% [46, 74, 87]. The territory of a city district can include green zones, parks, urban forests and even grasslands and croplands [62, 85, 87]. Sport and ornamental green lawns occupy up to 40% of the non-sealed territory [60]. Urban soils can have very different origin, from being part of what remains of natural ecosystems to completely artificial infrastructures. Urban ecosystems are characterized by a number of specific features and processes: soil compaction, sealing, pollution, adding organic composts and fertilizers in greenery practices [57, 66, 88, 92]. These specific conditions obviously contribute to urban soils' functions and biochemical processes, including soil respiration [37, 39, 89].

Urban soils are exposed to a wide variety of managements and disturbance, which provide conditions for significant differences of soil respiration. This constrains the spatial analysis of urban soils. Besides, relatively little information is available on the effect of land-use and management conditions, on the two major flux components of soil respiration: autotrophic respiration of root systems and root-associated organisms and heterotrophic respiration of free-leaving microorganisms in the soil (soil organic matter (SOM) – derived respiration) [14, 24]. Separation of total CO_2 efflux from soil into heterotrophic and autotrophic parts is crucial to understand C turnover and to predict the impact of land management on soil CO_2 emissions [52, 92]. Several methods are available to distinguish between autotrophic and heterotrophic respiration, both in laboratory and field conditions. The most frequently used are based on isotopic approaches, trenching and field segregation [11, 28, 65, 83]. The field segregation approach is the most cost-efficient, allowing for a large number of sampling points [22, 55], which is essential in heterogeneous urban areas.

The present study aimed to investigate the magnitude of soil respiration fluxes and the relative contribution of the autotrophic and heterotrophic components in urban soils under contrast land use types (green lawns, urban forest and cropland) within Moscow megalopolis.

2 Materials and Methods

2.1 Study Area

The sampling sites were located in the North of the Moscow city (N55°50'; E37°33') and represent adjacent green lawn, cropland and forest areas (in total 16 plots) [92]. The Moscow city is located in the Southern mixed forests vegetation subzone of a taiga-forest zone. The experimental area belongs to the most southern part of the Klinsko-Dmitrovskaya chine's slope. Relief is represented mainly by moraine hilly plain and moraine loamsare the main parent material. Zonal soddy-podzolic soils (Eutric Podzoluvisols) are the most widely distributed soil type in the area [18, 78, 92]. The climate in the Moscow city is humid continental with average July temperature 19.1 °C and average January temperature of −14.0 °C. In winter, temperature normally drops to approximately −10.0 °C, though there can be periods of warmth with temperature rising above 0.0 °C. The average number of days with temperature below zero varies from 151 to 197 with the clear tendency to decrease during the last decades. The average annual precipitation is close to 650 mm. Summer period lasts from mid-May to the beginning of September, whereas winter - from the beginning of November to the end of March, with snow cover starting around the beginning of November and melting generally at the beginning of April [63, 92].

2.2 Experimental Design

Three contrasting ecosystems were chosen within the territory of Russian State Agricultural University (RSAU) of Moscow megapolis, in adjacent areas: green lawns, which represent the highest percentage of non-sealed surface within the Moscow city areas were compared to the forest and croplands, which were chosen as less disturbed reference sites and which are also a quite representative surface (approximately 10% of the total area [10, 32, 92]. The lawns (six experimental plots) are dominated by meadow grass, red fescue, bent grass, daisy and ryegrass. Soil is represented by different sub-types of urban soils (Urban Technosols): urbanozems, replantozems and urban constructed soils [23, 69, 73]. Forest had five plots. This included a mixed forest with domination of pine, lime, birch, maple, oak, elm and larch species. Understorey vegetation was represented by hazelwort, gill, shield, fern and carex. Soddy-podzolic and peat-podzolic soils (Eutric and Orthic Podzoluvisols) dominate in the area [63, 92]. Cropland site was characterized by a four-field crop rotation: winter wheat-potatoes-barley- perennial grassland (vetch-oats mixture) under contrast management (tillage and non-tillage). Four experimental plots included wheat and potato both with tillage and non-tillage management. Soils are classified as agro-soddy-podzolic soils - zonal soils with a transformed profile and soil features (plough layer, soil compaction etc.) due to continuous use in agriculture [59, 92]. Each experimental plot was approximately 25 m^2. Within each plot three measuring points, each including three collars (control, autotrophic and heterotrophic respiration) were set-up.

2.3 Soil Respiration Partitioning and Measurements in Situ

The root exclusion technique was used to quantify the contribution of heterotrophic and autotrophic components to total *in situ* respiration [55, 61]. The procedure is based on the comparison of CO_2 fluxes from collars which are placed on (i) undisturbed soil (control, total respiration), (ii) reconstructed soil without roots and isolated with geo-textile material (1 micron mesh) to prevent further root growth inside the bag, enabling however water filtration (root exclusion), and (iii) reconstructed soil without roots and isolated with nylon net (1 cm mesh) which is permeable to both outside roots and water (disturbed control). As root removal in field induces soil disturbance a lag period occurs between experimental set-up and start of measurements. The first preparatory phase was hence started immediately after the snow melting and soil drying up (first decade of May). Soil was removed from holes of 35 cm in depth and 20 cm in diameter and the tissue bags made of two different materials were inserted in the hole. The soil samples were sieved (2 mm mesh sieve) and all the roots were carefully removed, afterwards soil was used to fill the space left by the hole, taking into account the vertical profile of soil horizons and sub-horizons [92]. Plastic collars were inserted on the top of disturbed and undisturbed soil at a depth of 4 cm (Fig. 1). Three collars were used for each treatment within each plot (giving 9 collars per plot).]

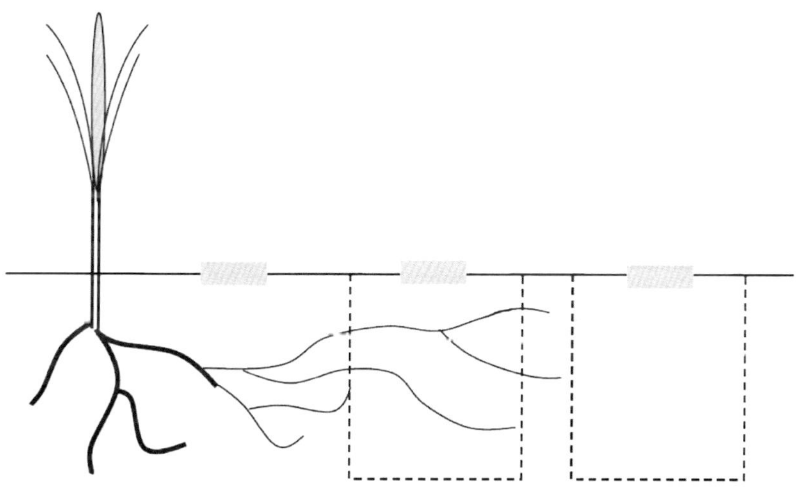

Fig. 1. Separation of soil for partitioning auto- and heterotrophic respiration in situ.

CO_2 flux measurements started in July in order to allow for sufficient root growth in the nylon bags. In order to perform comparative analysis of temporal dynamics in total soil respiration and its components throughout the vegetation season over the studied land-uses, flux measurements were made with a weekly frequency and ended at the end of September after the crops on the agricultural areas were harvested and further comparative analysis was not possible. In October the snow was again covering the land. The CO_2 flux was quantified using a Li-6400-XT system (LI-COR Biosciences, USA)), provided with a close dynamic chamber which was inserted on soil collars during the measurement.

With this experimental design the CO_2 flux measured from the 1 micron mesh bags is considered as the respiration associated with microbial decomposition of SOM, heterotrophic respiration (Rh*), whereas that measured from 1 cm mesh bags is assumed to be the sum of both heterotrophic and root-derived (autotrophic) respiration (Rh + Ra). The latter minus Rh* provides, autotrophic respiration (Ra). The CO_2 flux measured from collars over undisturbed soil represents the total soil respiration (Rs) in undisturbed conditions. The activity of sieving introduces disturbance which might in particular change the rate of heterotrophic respiration. Thus in order to quantify a more realistic efflux of CO_2 from microbial activity, Rh is calculated as Rs-Ra [22].

Soil temperature was measured (Checktemp thermometer, Hanna, Germany) for each chamber in 0–10 cm depth, 5 cm from the chamber edge by three replicates, at each sampling date. Air temperature and precipitation were obtained daily from Michelson weather station (55°50'9"N; 37°33'18"E) located at within 2 km distance from experimental sites [92].

2.4 Soil Chemical Properties and Microbial Biomass Carbon

For each experimental plot we analyzed soil pH and soil organic C (SOC) content as the basic soil features influencing soil respiration. Soil pH_{KCl} was measured by potentiometric analysis on KCl solutions (soil: KCl solution = 1: 2.5). SOC content was measured by dichromate oxidation [91]. As biological parameters related both to soil respiration and land management (SOM and disturbance) we also analyzed soil microbial biomass C (C_{mic}) in topsoil and subsoil horizons (averaged depth 0–30 cm and 30–60 cm respectively). C_{mic} was determined by the substrate-induced respiration (SIR) [3, 4, 81]. Samples were preincubated for 7 days at 22 °C and 55% of water holding capacity [1]. Thereafter subsamples (2 g) in 4 replicas were taken for respiration measurements from pre-incubated soils, and soil was incubated in air tight vials (15 ml volume) with 2 μl glucose solution (10 mg glucose g^{-1} soil) for 3–5 h at 22 °C [15] after which headspace gas samples were collected and CO_2 concentration was determined using a Chrom-5 model gas chromatograph equipped with a thermal conductivity detector. For each soil the optimal incubation to have the maximum peak was previously determined in a preliminary run. Microbial biomass C (μg C g^{-1} soil) was estimated from SIR according to the equation: C_{mic} (μg C g^{-1}) = SIR (μl CO_2 g^{-1} h^{-1}) × 40.04 + 0.37 as described in Anderson andDomsch (1978) [2].

2.5 Statistical Analysis

Descriptive statistics and correlation matrixes were used to process the data and to investigate spatial and temporal variability of soil total, autotrophic and heterotrophic respiration within and between urban, cropland and forest sites, as well as main investigated soil parameters. In order to check the assumptions of normality for the analysis of variance (ANOVA) QQ-plots were built up. The homogeneity of variance was checked using Levene's test. The independence of the data was appropriately addressed by the sampling design. Main-effect and factorial ANOVA was implemented to analyze the impact of the «land-use» and «time» factors on soil respiration. Predictive power of ANOVA was characterized by determination coefficients R^2 and R^2_{adj}. An F-protected LSD test was used to check for significance of differences between the groups.

3 Results

3.1 Soil Chemical Properties and Microbial Biomass Carbon

Considering the close proximity of the analyzed sites climatic conditions were comparable and the lithological origin was the same for all the investigated soils [92]. We studied the spatial variability and profile distribution of the two parameters with known substantial influence on soil respiration −pH and SOC content. Urban forest soils, which were less exposed to anthropic influence and disturbance, showed the lowest pH (comparable and close to zonal soils) and the highest average SOC in the top soil, both abruptly decreasing in eluvial horizon, below 30 cm depth. In cropland and green lawns, where anthropogenic disturbance was much more relevant, the SOC content was lower than in the forest site, in particular in the topsoil (Fig. 2). When considering the whole soil profile down to 100 cm, however, results varied significantly with the analyzed site. The cropland soil, exposed to both liming and tillage, showed lower average topsoil SOC and higher topsoil pHvalues in comparison to the forest site. Green lawns, where the strongest transformation and manipulation occurred, showed the highest variability of SOC and pH with pH values ranging from 3.8 to 6.0 and average SOC over 100 cm depth varying from 0.5 to 2.1%.

Microbial biomass C was significantly higher in the topsoil, compared with the subsoil in all the three analyzed systems. C_{mic} content in the forest and lawn soils were comparable, whereas C_{mic} in the cropland soil was on average lower (Fig. 3).

3.2 In Situ Measurements of Soil Respiration and Its Components

During the period of study there were significant variations in soil temperature with the highest daily average value recorded at the end of July, 22 °C and the lowest, 15 °C, in September. Significant differences of total soil respiration Rs were found among land-use types In order to characterize Ra and Rh by root exclusion technique two assumptions are made. Firstly, the procedure of excluding roots disturbs the soil so that to compared the different component of soil respiration a "disturbed control is needed". Secondly, it is necessary to leave a lag time between treatment preparation and starting of respiration measurements during which plant roots are again able to recolonize the "disturbed control" so that the autotrophic contribution in disturbed and undisturbed site can be considered to be comparable. Data show a very good positive correlation ($r = 0.78$, $p < 0.05$) between both measurements (Fig. 4). This result provides evidence that an optimal level of recovery from disturbance in the treated plots was obtained. Rs, Ra and Rh, measured in the observed forest, cropland and lawn sites were significantly different (Fig. 5). The mean Rs value obtained for lawns was significantly

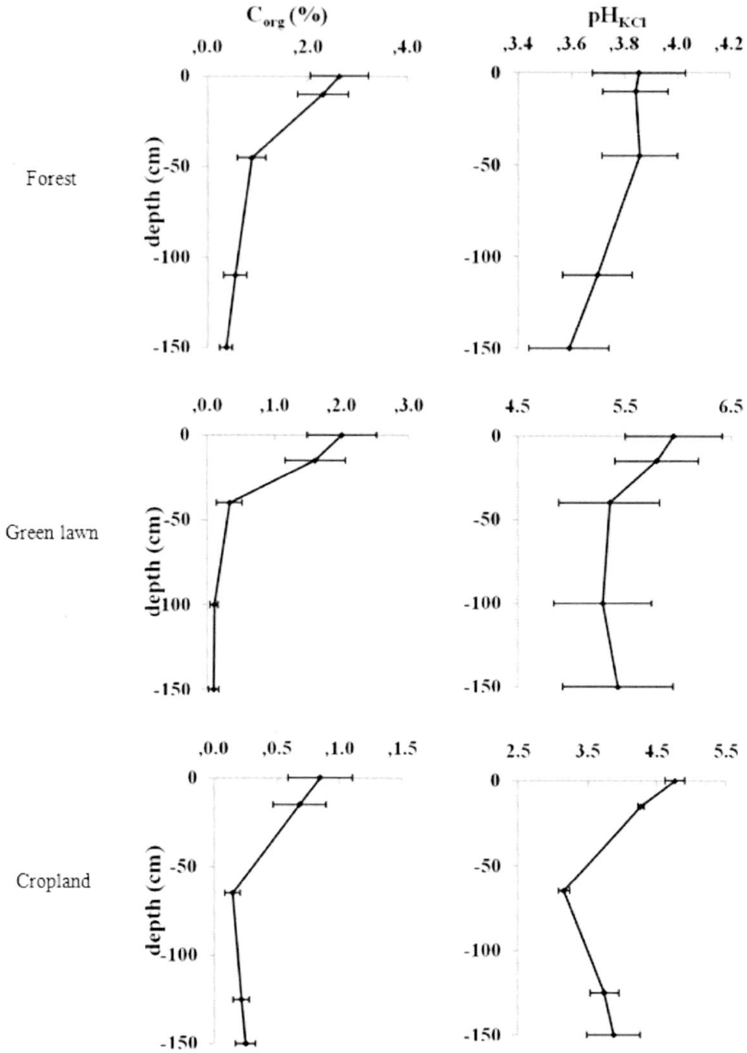

Fig. 2. Soil profile distribution of C_{org} and pH_{KCl} in the observation sites (mean and stadard error values, averaged per each site and depth).

higher than at the other sites (F-protected LSD-test, $p < 0.05$). Rs in the forest site was on average 20% higher than one measured in croplands, but the difference was not statistically significant. The highest average Ra value was also obtained for lawns, and it was 40% higher than Ra measured in the forest site and was twice the value recorded in the cropland. Rh was significantly higher than Ra in all the studied plots and during the whole observation period. The highest average Rh was obtained for urban lawns, and was twice the CO_2 efflux measured in forest and cropland sites, which instead were comparable (Fig. 5). Although Rs and its components were quite different among land

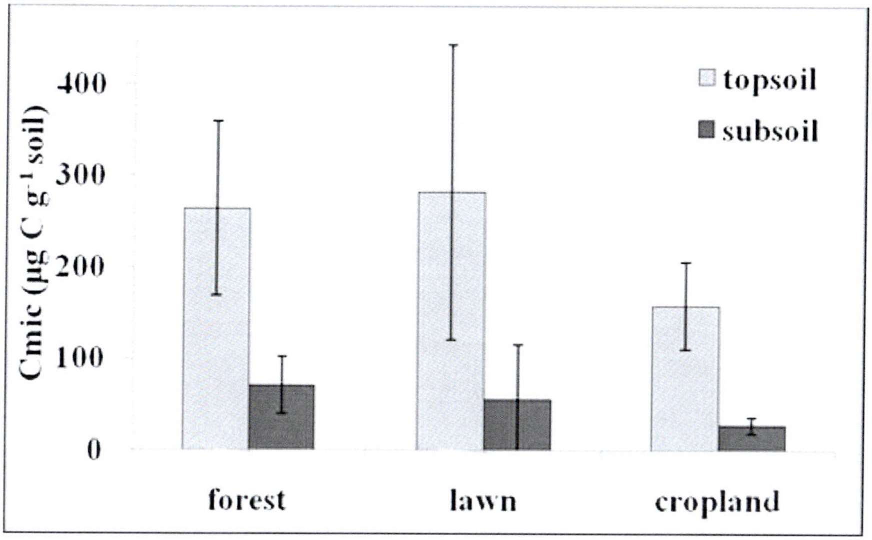

Fig. 3. Average microbial biomass C (μg C g^{-1}) in forest, cropland and lawns soils (μ ± SE)

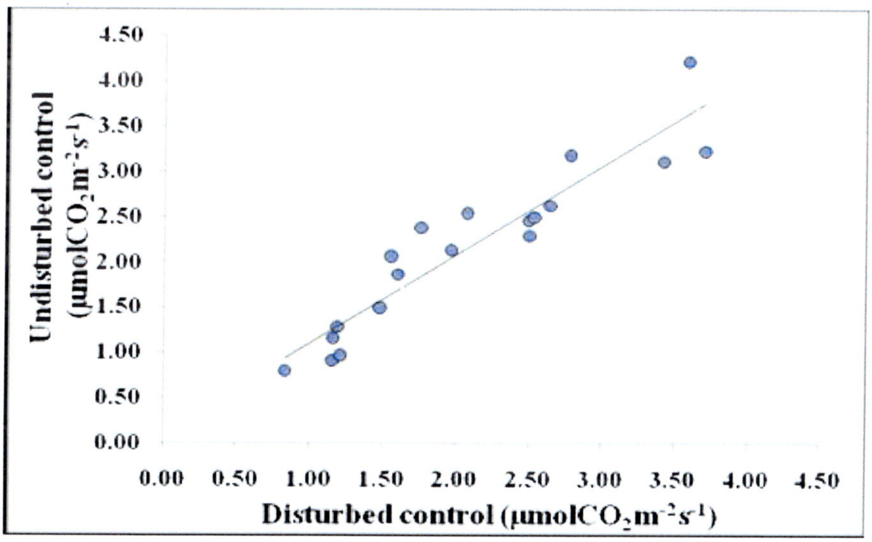

Fig. 4. Plot of CO_2 fluxes from disturbed vs undisturbed control replicates.

uses, the relative contribution of Ra to Rs showed similar results for all the sites, giving on average 25% for urban lawns, 27% for croplands and 28% for urban forest (Table 1). In terms of temporal variations Rs, Ra and Rh showed different trends. In fact, Ra and, at a minor extent, Rs increased along the observation period, with the

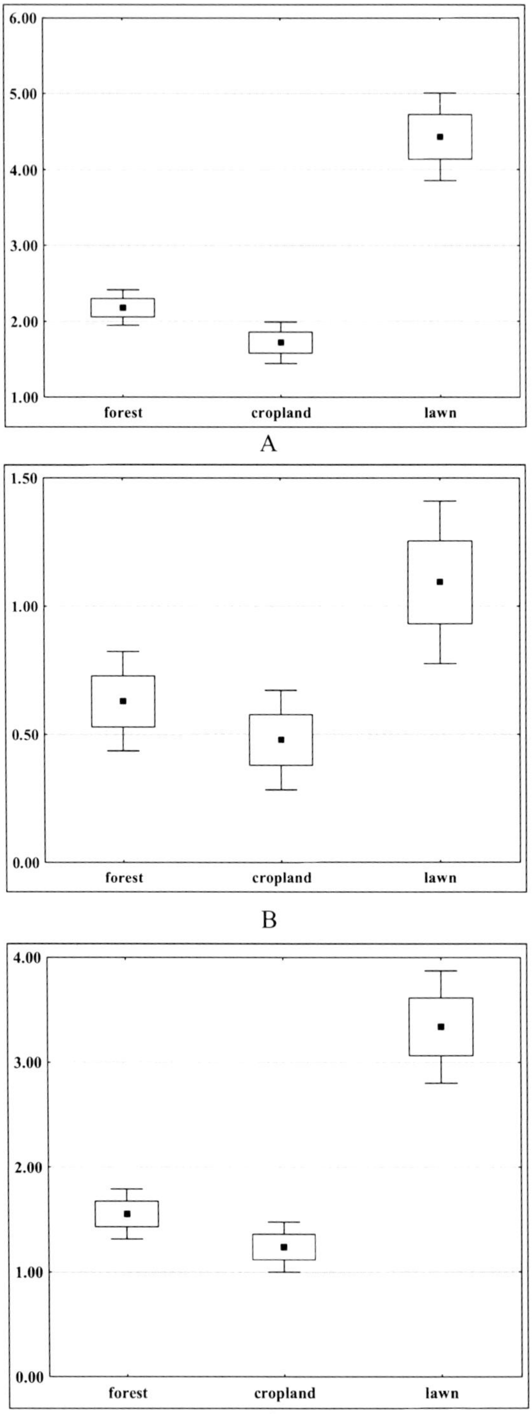

Fig. 5. Mean, standard error (SE) and confident interval (1.96 × SE) values for Rs (A), Ra (B), Rh (C) (μmolCO_2 m^{-2} s^{-1}) in the forest, cropland, and urban lawns study sites

Table 1. Chemical features and total (Rs), autotrophic (Ra) and heterotrophic (Rh) respiration and the contribution of autotrophic component to total respiration (RaC) for different sites

	Urban forest	Green lawns	Cropland
Soil chemical features, averaged for 0–100 cm			
C_{tot} (%)	1.09 ± 0.23	1.34 ± 0.47	1.05 ± 0.33
pH_{KCl}	4.68 ± 0.43	6.40 ± 0.54	5.10 ± 0.13
Soil respiration and it's components (*in situ* measurements)			
Rs (μmol CO_2 $m^{-2}s^{-1}$)	2.18 ± 0.70	4.43 ± 1.91	1.75 ± 0.15
Ra (μmol CO_2 $m^{-2}s^{-1}$)	0.62 ± 0.19	1.09 ± 0.45	0.45 ± 0.13
Rh (μmol CO_2 $m^{-2}s^{-1}$)	1.55 ± 0.25	3.39 ± 1.78	1.38 ± 0.15
RaC (%)	28 ± 6	25 ± 10	27 ± 6

steeper increase for Ra (Fig. 6A and B). On the contrary, no clear temporal trend was observed for Rh (Fig. 6C). The contribution of Ra to Rs was continuously increasing over the observation period for all the sites from an average of 16–18% at the beginning to about 30–35% at the end of experiment.

4 Discussion

4.1 Soil Respiration and Its Components in Urban, Forest and Cropland Soils

Land-use is the most important factor affecting soil C balance and thus is incorporated in the majority of global and regional C models [43, 67, 79]. Our experimental design allowed comparing soil respiration in sites characterized by very different land-use but in similar climatic conditions and with same parental material [92]. Soil respiration is a measure of both microbial activity and root respiration. The latter is strongly driven by the phonological cycle of the plants and at a minor extent is also linked to stress conditions for the plant [44, 50, 54]. Microbial respiration is generally considered proportional to the amount of soil organic matter and microbial biomass present in the soil [1, 49]. Higher respiration rates per unit of microbial biomass and/or per unit of SOC, are often reported for soils presenting unfavorable conditions for microorganisms [6, 17, 33].

We observed the highest mean soil respiration rates in the most disturbed urban soils, the green lawn sites, compared with soils of the agricultural and forest sites. This finding is in accordance with few available similar studies on urban soils in Arizona and Baltimore (USA) [39, 41] and Shanghai and Nanjing in China [80]. The drivers influencing high observed respiration of urban soils respiration are still poorly understood. Some authors explain this result with the higher SOC content usually reported for urban soils, often amended with artificial organic substrates (compost and sewage materials), during soil reclaiming and greenery work [68, 92]. These substrates are rich in labile form of organic C which can be more easily mineralized by microorganisms [7, 57, 85]. Another explanation links the higher urban soil respiration rates to higher average soil and air temperatures encountered in urban areas as a

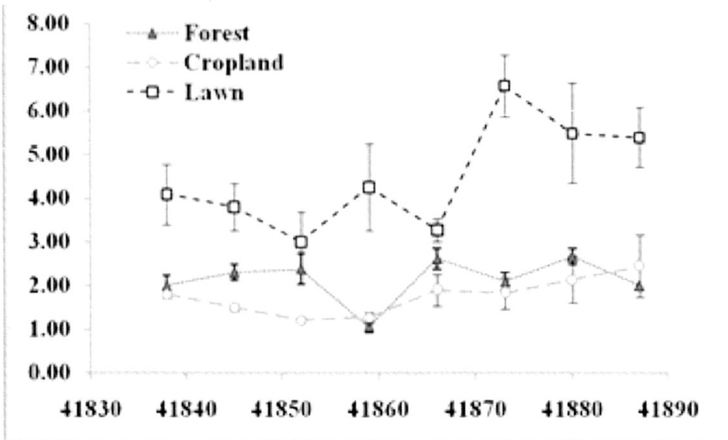

A

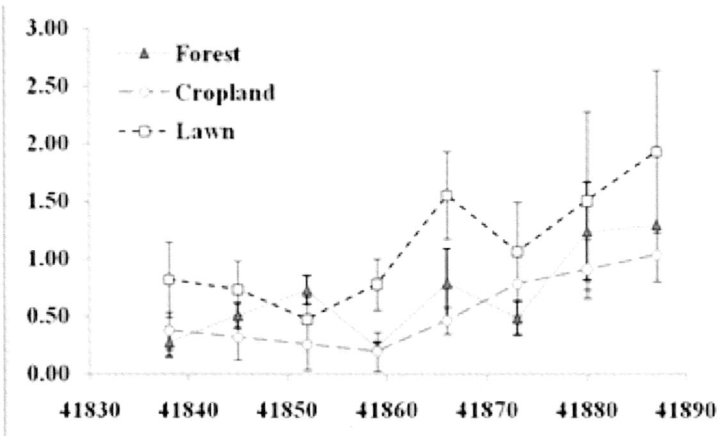

B

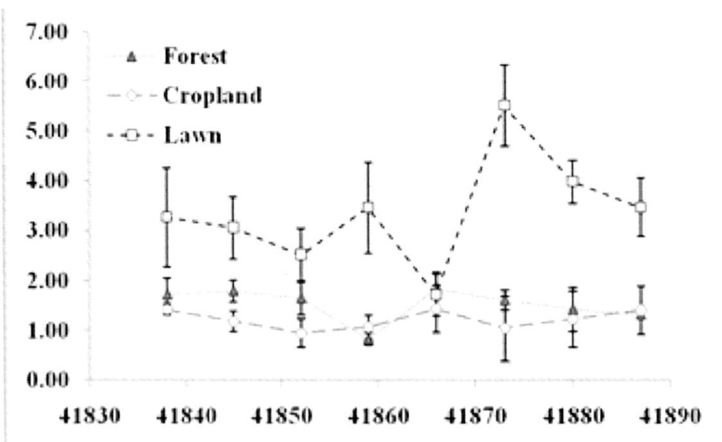

C

Fig. 6. Temporal dynamic of Rs (A), Ra (B) and Rh (C) (all in μmol CO_2 $m^{-2}s^{-1}$) calculated for the whole monitoring period and averaged for lawn, cropland and forest soils.

consequence of the heat island effect [77]. Others attribute higher respiration rates to the alteration of soil physical properties and soil pollution. The effects of this latter factor on soil respiration is however not well defined, as in some cases pollution has been related to increasing rates of soil respiration [9, 40, 86] and in other cases - to its reduction [5, 13, 64]. Results seem to vary significantly depending on the contaminant type and its concentration.

In our study we didn't find statistically significant differences in SOC contents among most of the sites. Temperature effect was also negligible. Thus we could assume that the level of anthropogenic disturbance might influence microbial conditions, activity and soil respiration. However, being the Rs the sum of both Ra and Rh component we cannot exclude that differences in soil respiration might not be due to difference in autotrophic respiration. Our experimental design allowed analyzing the contribution of both components. Data showed that for all land-use types microbial respiration was the dominant component of soil respiration, accounting on average for never less than 72% of total respiration (Rs). A direct comparison with published data is not straightforward due to the high variability of the reported results and the lack of similar data for urban soils. Different sources report a contribution of autotrophic respiration to total *in situ* soil respiration which goes from 10% to 90% depending on vegetation type and season of the year [8, 24, 28]. There are also evidences proving higher contribution of root respiration in forests than in grasslands [71] and a higher heterotrophic contribution in croplands, compared with natural areas [12]. The only available study where root and microbial respiration was partitioned in urban green lawns [56] reported a root contribution to Rs of 26%, hence a heterotrophic respiration contribution of 74%, very close to our findings.

Higher microbial activity in urban lawns is likely stimulated by additional irrigation and fertilization [25]. Stressful environmental conditions also might favor dissipative metabolism vs. immobilization of C into microbial biomass [58]. Anderson and Domsch (1978; 1986) [2, 4] have expressed this concept using an indicator of dissipative microbial metabolism, the metabolic quotient. Following this concept and substituting basal respiration by *in situ* Rh we can calculate the qCO_2 of our soils as Rh/C_{mic}. What we obtain is that the highest $Rh/C_{mic}$is found in the green lawns, followed by cropland and forest soils (1.53; 0.73 and 0.58 respectively). This result is in accordance with a trend of increasing disturbance and increased CO_2 loss per unit of microbial C going from forest sites to urban lawns. This finding is in accordance with soil studies reporting higher respiration ratesfor disturbed soils [9, 40].

4.2 Temporal Trends of Soil Respiration

The observed temporal trends of total soil respiration were similar for all the plots, with maximum Rs rates occurring at the end of August - beginning of September and the minimum- at the end of July [92]. This temporal trend is similar to what reported in literature for temperate [45, 47, 53] and Mediterranean climate [72]. However, July in Moscow is less dry and September is less warm than in Mediterranean countries, and this might justify why the difference between summer and autumn effluxes was not so pronounced and resulted not statistically significant. When, however, the temporal dynamics of heterotrophic and autotrophic respiration (Ra and Rh) were analyzed separately it was possible to appreciate some significant temporal trends and differences between Ra and Rh. We evidenced as while a temporal trend was evidenced for Ra, Rh did not show any clear temporal trend. It might be possible that the two factors, which strongly influence microbial activity and CO_2 emissions, i.e. temperature and soil water content, counterbalanced each other. July was in fact characterized by higher temperatures but drier conditions, while August-September by wetter conditions and lower temperatures. Several studies have shown the synergic positive effect of these two environmental variable on microbial respiration rates [15, 75] confirm this assumption. On the contrary, autotrophic respiration clearly showed an increase along the season, which could be very well justified with the phonological cycle of the plant communities, leading to a progressive increasing of root biomass along the vegetative season from July to September. Several studies have reported the more relevant role of physiological over abiotic factors in controlling root respiration [19, 29, 51].

5 Conclusion

C stocks and fluxes of urban soils lack attention and our understanding of urban soil respiration remain very limited and uncertain. In our research we demonstrated that respiration of urban soil was significantly higher than in adjacent cropland (managed) and forest (natural) soils. We showed that the highly heterogeneous urban environment, both in terms of management practices and anthropogenic disturbance results in high spatial heterogeneity of soil respiration, with coefficients of variance higher than 60% and 2–3 times higher than reported in forest and cropland sites, highlighting the difficulty of characterizing this CO_2 source [92]. We also proved that the microbial community had a dominant role in soil respiration in the urban environments, and that the heterotrophic respiration was enhanced by disturbance and atrophic pressure.

Although our results represent a preliminary study they contribute to implement our understanding of CO_2 emissions from urban soils. The contribution of urban soils to regional C balance will be progressively more important in the future when urbanization and pollution will be among the most important factors affecting soil quality and health.

Acknowledgements. The project was supported by Russian Science Foundation project № 17-77-20046. The authors thank Pavel Lakeev, Irma Elvira Ade and Dmitry Gusev for assistance with the field measurement as well as Dr. Julia Kurbatova for valuable suggestions and useful comments.

References

1. Ananyeva, N.D., Susyan, E.A., Chernova, O.V., Wirth, S.: Microbial respiration activities of soils from different climatic regions of European Russia. Eur. J. Soil Biol. **44**, 147–157 (2008)
2. Anderson, J.P., Domsch, K.H.: A physiological method for the quantitative measurement of microbial biomass in soils. Soil Biol. Biochem. **10**, 215–221 (1978)

3. Anderson, T.H., Domsch, K.H.: Determination of ecophysiological maintenance requirements of soil micro-organisms in a dormant state. Biol. Fert. Soils **1**, 81–89 (1985)
4. Anderson, T.H., Domsch, K.H.: C links between microbial biomass and soil organic matter. In: Megusar, F., Gantar, M. (eds.) Perspectives in Microbial Ecology, pp. 467–471. Slovene Society for Microbiology, Ljubljana (1986)
5. Barajas-Aceves, M.: Comparison of different microbial biomass and activity measurement methods in metal contaminated soils. Bioresour. Technol. **96**, 1405–1414 (2005)
6. Bastida, F., Zsolnay, A., Hernandez, T., Garcia, C.: Past, present and future of soil quality indices: a biological perspective. Geoderma **147**, 159–171 (2008)
7. Beesley, L.: C storage and fluxes in existing and newly created urban soils. J. Environ. Manag. **104**, 158–165 (2012)
8. BelelliMarchesini, L., Papale, D., Reichstein, M., Vuichard, N., Tchebakova, N., Valentini, R.: C balance assessment of a natural steppe of southern Siberia by multiple constraint approach. Biogeosciences **4**, 581–595 (2007)
9. Blagodatskaya, E.V., Pampura, T.V., Dem'yanova, E.G., Myakshina, T.N.: Effect of lead on growth characteristics of microorganisms in soil and rhizosphere of Dactylisglomerata. Eurasian Soil Sci. **39**(6), 635–660 (2006)
10. Bochin, L.A. (ed.): Report of the Environment in Moscow in 2007. Department for nature use and environment protection of Moscow city, Moscow (2008)
11. Bond-Lamberty, A., Thomson, A.: A global database of soil respiration data. Biogeosciences **7**, 1915–1926 (2010)
12. Buyanovsky, G.A., Wagner, G.H.: C cycling in cultivated land and its global significance. Glob. Change Biol. **4**(2), 131–141 (1998)
13. Castaldi, S., Rutigliano, F.A., Virzo de Santo, A.: Suitability of soil microbial parameters as indicators of heavy metal pollution. Water Air Soil Poll. **158**, 21–35 (2004)
14. Chapin III, F.S., Woodwell, G.M., Randerson, J.T., et al.: Reconciling C-cycle concepts, terminology, and methods. Ecosystems **9**, 1041–1050 (2006)
15. Creamer, R.E., Schulte, R.P.O., Stone, D., et al.: Measuring basal soil respiration across Europe: do incubation temperature and incubation period matter? Ecol. Ind. **36**, 409–418 (2014)
16. Cruvinel, E.F., Bustamante, M.M.C., Kozovits, A., Zepp, R.G.: Soil emission of NO, N_2O and CO_2 from croplands in the savanna region of central Brazil. AgrEcosyst. Environ. **144**, 29–40 (2011)
17. Dilly, O., Blume, H.P., Sehy, U., Jimenez, M., Munich, J.C.: Variation of stabilized, microbial and biologically active C and nitrogen soil under contrasting land use and agricultural management practices. Chemosphere **52**, 557–569 (2003)
18. FAO: Soils Map of the World: Revised Legend. Food and Agriculture Organization of the United Nations, Rome (1988)
19. Ferrea, C., Zenone, T., Comolli, R., Seufert, G.: Estimating heterotrophic and autotrophic soil respiration in a semi-natural forest of Lombardy, Italy. Pedobiologia **55**, 285–294 (2012)
20. Fromin, N., Porte, B., Lensi, R., Hamelin, J., Domenach, A.-M., Buatois, B., Roggy, J.-C.: Spatial variability of the functional stability of microbial respiration process: a microcosm study using tropic forest soil. J. Soils Sed. **12**, 1030–1039 (2012)
21. Galford, G.L., Melillo, J., Mustard, J.F., Cerri, C.E.P., Cerri, C.C.: The Amazon frontier of land-use change: croplands and consequences for greenhouse gas emissions. Earth Interact. **14**, 1–20 (2010)
22. Gavrichkova, O.: Drivers of soil respiration of root and microbial origin in grasslands. Doctorate thesis. Tuscia University. Viterbo (2010)
23. Gerasimova, M.I., Stroganova, M.N., Mozharova, N.V., Prokofieva, T.V.: Urban Soils. Oykumena, Smolensk (2003)

24. Gomes-Casanovas, N., Matamala, R., Cook, D.R., Gonzalez-Meler, M.A.: Net ecosystem exchange modifies the relationship between the autotrophic and heterotrophic components of soil respiration with abiotic factors in prairie grasslands. Glob. Change Biol. **18**, 2532–2545 (2012)
25. Groffman, P.M., Law, N.L., Belt, K.T., Band, L.E., Fisher, G.T.: Nitrogen fluxes and retention in urban watershed ecosystems. Ecosystems **7**, 393–403 (2004)
26. Guo, L.B., Gifford, R.M.: Soil C stocks and land use change: a meta-analysis. Glob. Change Biol. **8**, 345–360 (2002)
27. Hamilton, J.G., DeLucia, E.H., George, K., Naidu, S.L., Finzi, A.C., Schlesinger, W.H.: Forest C balance under elevated CO_2. Oecologia **131**, 250–260 (2002)
28. Hanson, P.J., Edwards, N.T., Garten, C.T., Andrews, J.A.: Separating root and soil microbial contributions to soil respiration: a review of methods and observations. Biogeochemistry **48**, 115–146 (2000)
29. Heinemeyer, A., Totrorellal, D., Petrovicova, B., Gelsomino, A.: Partitioning of soil CO_2 flux components in a temperate grassland ecosystem. Eur. J. Soil Sci. **63**(2), 249–260 (2012)
30. Houghton, R.A.: Why are estimates of the terrestrial C balance so different? Glob. Change Biol. **9**, 500–509 (2003)
31. Hughes, R.F., Kauffman, J.B., Jaramillo, V.J.: Biomass, C, and nutrient dynamics of secondary forests in a humid tropical region of Mexico. Ecology **80**, 1892–1907 (1999)
32. Il'ina, I.N. (ed.): Ecological Atlas of Moscow. ABF, Moscow (2000)
33. Insam, H., Hutchinson, T.C., Reber, H.H.: Effects of heavy metal stress on the metabolic quotient of soil microflora. Soil Biol. Biochem. **28**, 691–694 (1996)
34. Metz, B., Davidson, O., de Coninck, C.H., Loos, M., Meyer, L.A. (eds.) IPCC (2005): IPCC special report on C dioxide capture and storage', prepared by Working Group III of the Intergovernmental Panel on Climate Change. CambridgeUniversity Press, Cambridge, United Kingdom and New York, NY, USA, p. 442 (2005)
35. Stocker, T.F., Qin, D., Plattner, G.K., Tignor, M., Allen, S.K., Boschung, J., Nauels, A., Xia, Y., Bex, V., Midgley, P.M. (eds.): IPCC (2013): Climate Change 2013: The Physical Science Basis. Contribution of Working Group I to the Fifth Assessment Report of the Intergovernmental Panel on Climate Change. Cambridge University Press, Cambridge, United Kingdom and New York, NY, USA, 1535 p. (2013)
36. Islam, K.R., Weil, R.R.: Land use effects on soil quality in a tropical forest ecosystem of Bangladesh. Agric. Ecosyst. Environ. **79**, 9–16 (2000)
37. Ivashchenko, K.V., Ananyeva, N.D., Vasenev, V.I., Kudeyarov, V.N., Valentini, R.: Biomass and respiration activity of soil microorganisms in anthropogenically transformed ecosystems (Moscow region). Eurasian Soil Sci. **47**(9), 892–903 (2014)
38. Janzen, H.H.: C cycling in earth systems - a soil science perspective. Agric. Ecosyst. Environ. **104**, 399–417 (2004)
39. Kaye, J.P., McCulley, R.L., Burkez, I.C.: C fluxes, nitrogen cycling, and soil microbial communities in adjacent urban, native and agricultural ecosystems. Glob. Change Biol. **11**, 575–587 (2005)
40. Khan, M., Scullion, J.: Effects of metal (Cd, Cu, Ni, Pb or Zn) enrichment of sewage-sludge on soil micro-organisms and their activities. Appl. Soil. Ecol. **20**(2), 145–155 (2002)
41. Koerner, B., Klopatek, J.: Anthropogenic and natural CO2 emission sources in an arid urban environment. Environ. Pollut. **116**, 45–51 (2002)
42. Koechy, M., Hiederer, R., Freibauer, A.: Global distribution of soil organic C - Part 1: masses and frequency distributions of SOC stocks for the tropics, permafrost region, wetlands, and the world. Soil **1**, 351–356 (2015)
43. Komarov, A.S., Chertov, O.G., Zudin, S.L., et al.: EFIMOD 2—a model of growth and elements cycling of boreal forest ecosystems. Ecol. Model. **170**, 373–392 (2003)

44. Kozlowski, T.T., Pallardy, S.G.: Acclimation and adaptive responses of woody plants to environmental stresses. Bot. Rev. **68**(2), 270–334 (2002)
45. Kudeyarov, V.N., Zavarzin, G.A., Blagodatsky, S.A., et al.: C Pools and Fluxes in Terrestrial Ecosystems of Russia. Nauka, Moscow (2007)
46. Kurbatova, A.S., Bashkin, V.N., Barannikova, Yu.A., et al.: Ecological Functions of Urban Soils. Smolensk–Moscow (2004)
47. Kurganova, I.N., De Gerenuy, V.L., Rozanova, L., Sapronov, D., Myakshina, T., Kudeyarov, V.: Annual and seasonal CO_2 fluxes from Russian southern taiga soils. Tellus Ser. B: Chem. Phys. Meteorol. **55**(2), 338–344 (2003)
48. Kurganova, I.N., Yermolaev, A.M., Lopes de Gerenyu, V.O., Larionova, A.A., Kuzyakov, Y., Keller, T., Lange, S.: C balance in the soils of abandoned lands in Moscow region. Eurasian Soil Sci. **40**(1), 51–58 (2007)
49. Kuzyakov, Y.: Sources of CO_2 efflux from soil and review of partitioning methods (review). Soil Biol. Biochem. **38**, 425–448 (2006)
50. Kuzyakov, Y., Domanski, G.: C input by plants into soil. Review. J. Plant Nutr. Soil Sci. **163**, 421–431 (2000)
51. Kuzyakov, Y., Gavrichkova, O.: Review: time lag between photosynthesis and C dioxide efflux from soil: a review of mechanisms and controls. Glob. Change Biol. **16**, 3386–3406 (2010)
52. Kuzyakov, Y., Larionova, A.A.: Root and rhizomicrobial respiration: a review of approaches to estimate respiration by autotrophic and heterotrophic organisms in soil. J. Plant Nutr. Soil Sci. **168**(4), 503–520 (2005)
53. Larionova, A.A., Rozanova, L.N., Demkina, T.S., Evdokimov, I.V., Blagodatskii, S.A.: Annual emission of CO_2 from gray forest soil. Eurasian Soil Sci. **34**(1), 61–68 (2001)
54. Lavinge, M.B., Foster, R.J., Goodine, G.: Seasonal and annual changes in soil respiration in relation to soil temperature, water potential and trenching. Tree Physiol. **24**, 415–424 (2004)
55. Leake, J.R., Johnso, D., Donnelly, D.P., Muckle, G.E., Boddy, L., Read, D.J.: Networks of power and influence: The role of mycorrhizal mycelium in controlling plant communities and agroecosystem functioning. Can. J. Bot.- Revue Canadienne De Botanique **82**, 1016–1045 (2004)
56. Li, X., Zeng, W., Li, J., Huang, X., Zhong, X., Yang, Y.: Effects of short-term cold-air outbreak on soil respiration and its components of subtropical urban green spaces. Shengtai Xuebao/Acta Ecologica Sinica **31**(19), 5728–5738 (2011)
57. Lorenz, K., Lal, R.: Biogeochemical C and N cycles in urban soils. Environ. Int. **35**, 1–8 (2009)
58. Lysak, L.V., Lapygina, E.V., Konova, I.A., Zvyagintsev, D.G.: Population density and taxonomic composition of bacterial nanoforms in soils of Russia. Eurasian Soil Sci. **43**(7), 765–770 (2010)
59. Mazirov, M.A., Safonov, A.F.: Long-term field experiment in RSAU-MTAA: main points and development stages. Izvestya MTAA **2**, 66–75 (2010)
60. Milesi, C., Running, S.W.: Mapping and modeling the biogeochemical cycling of turf grasses in the United States. Environ. Manag. **36**, 426–438 (2005)
61. Moyano, F., Kutsch, W., Rebmann, C.: Soil respiration fluxes in relation to photosynthetic activity in broad-leaf and needle-leaf forest stands. Agric. For. Manag. **48**, 135–143 (2008)
62. Naizheng, X., Hongying, L., Feng, W., Yiping, Z.: Urban expanding pattern and soil organic, inorganic C distribution in Shanghai China. Environ. Earth Sci. **66**(4), 1233–1238 (2012)
63. Naumov, V.D. (ed.): 145 years to forest experimental station of RSAU-MTAA. RSAU-MTAA, Moscow (2009)

64. Nwachukwu, O.I., Pulford, I.D.: Microbial respiration as an indication of metal toxicity in contaminated organic materials and soil. J. Hazard. Mater. **185**, 1140–1147 (2011)
65. Pataki, D.E., Ellsworth, D.S., Evans, R., et al.: Tracing changes in ecosystem function under elevated C dioxide conditions. Bioscience **53**(9), 805–818 (2003)
66. Pickett, S.T.A., Cadenasso, M.L., Grove, J.M.: Urban ecological systems: scientific foundations and a decade of progress. J. Environ. Manag. **92**, 331–362 (2011)
67. Post, W.M., Kwon, K.C.: Soil C sequestration and land-use changes: processes and potential. Glob. Change Biol. **6**, 317–327 (2000)
68. Pouyat, R., Groffman, P., Yesilonis, I., Hernandez, L.: Soil C pools and fluxes in urban ecosystems. Environ. Pollut. **116**, S107–S118 (2002)
69. Prokofyeva, T.V., Martynenko, I.A., Ivannikov, F.A.: Classification of Moscow soils and parent materials and its possible inclusions in the classification system of Russian soils. Eurasian Soil Sci. **44**, 561–571 (2011)
70. Raich, J.W., Potter, C.S., Bhagawati, D.: Interannual variability in global respiration, 1980–94. Glob. Change Biol. **8**, 800–812 (2002)
71. Raich, J.W., Tufekcioglu, A.: Vegetation and soil respiration: correlations and controls. Biogeochemistry **48**, 71–90 (2000)
72. Rey, A., Pegorado, E., Tedeschi, V., De Parri, I., Jarvis, P.G., Valentini, R.: Annual variation in soil respiration and its components in a coppice oak forest in Central Italy. Glob. Change Biol. **8**, 851–866 (2002)
73. Rossiter, D.G.: Classification of urban and industrial soils in the world reference base for soil resources. J. Soils Sed. **7**, 96–100 (2007)
74. Scalenghe, R., Marsan, F.A.: The anthropogenic sealing of soil in urban areas. Landscape Urban Plan. **90**, 1–10 (2009)
75. Schaufler, G., Kitzler, B., Schindlbacher, A., Skiba, U., Sutton, M.A., Zechmeister-Boltenstern, S.: Greenhouse gas emissions from European soils under different land use: effects of soil moisture and temperature. Eur. J. Soil Sci. **61**(5), 683–696 (2010)
76. Schulze, E.D.: Biological control of the terrestrial C sink. Biogeosciences **2**, 147–166 (2006)
77. Shen, Z., Shi, B., Jiang, H.: The temperature dependence of soil organic matter decomposition and CO_2 efflux: a review. Shengtai Xuebao/Acta Ecologica Sinica **33**(10), 3011–3019 (2013)
78. Shishov, V. (ed.): Soils of Moscow Region and Their Use. Dokuchaev Soil Science Institute, Moscow (2002)
79. Sitch, S., Smit, B., Prentice, I.C., Arneth, A., et al.: Evaluation of ecosystem dynamics, plant geography and terrestrial C cycling in the LPJ dynamic global vegetation model. Glob. Change Biol. **9**, 161–185 (2003)
80. Sun, Q., Fang, H.L., Liang, J., Qian, X.W., Liu, M.D., Zhang, Q.F., Hao, R.J., Hao, G.J.: Soil respiration characteristics of typical urban lawns in Shanghai. Chin. J. Ecol. **28**(8), 1572–1578 (2009)
81. Susyan, E.A., Ananyeva, N.D., Blagodatskaya, E.V.: The antibiotic-aided distinguishing of fungal and bacterial substrate-induced respiration in various soil ecosystems. Microbiology **74**, 336–342 (2005)
82. Swift, R.S.: Sequestration of C by soil. Soil Sci. **166**, 858–871 (2001)
83. Taneva, L., Gonzalez-Meler, M.: Distinct patterns in the diurnal and seasonal variability in four components of soil respiration in a temperate forest under free-air CO_2 enrichment. Biogeosciences **8**(10), 3077–3092 (2011)
84. Thuille, A., Buchmann, N., Schulze, E.D.: C stocks and soil respiration rates during deforestation, grassland use and subsequent Norway spruce afforestation in the Southern Alps Italy. Tree Physiol. **20**(13), 849–857 (2000)

85. Vasenev, V.I., Stoorvogel, J.J., Vasenev, I.I., Valentini, R.: How to map soil organic stocks in highly urbanized region? Geoderma **226-227**, 103–115 (2014)
86. Vasenev, V.I., Ananyeva, N.D., Ivashchenko, K.V.: Influence of contaminants (oil products and heavy metals) on microbiological activity of urban constructed soils. Russ. J. Ecol. **6**, 475–483 (2013a)
87. Vasenev, V.I., Prokofieva, T.V., Makarov, O.A.: Development of the approach to assess soil organic C stocks in megapolis and small settlement. Eurasian Soil Sci. **6**, 1–12 (2013b)
88. Vasenev, V.I., Stoorvogel, J.J., Vasenev, I.I.: Urban soil organic C and its spatial heterogeneity in comparison with natural and agricultural areas in the Moscow region. Catena 107, 96–102 (2013c)
89. Vasenev, V.I., Ananyeva, N.D., Makarov, O.A.: Specific features of the ecological functioning of urban soils in Moscow and Moscow oblast. Eurasian Soil Sci. **45**, 194–205 (2012)
90. Vasenev, V., Cataldi, S., Vizirskaya, M., Ananyeva, N., Ivashchenko, K., Valenini, R., Vasenev, I.: Root and microbial respiration from urban, agricultural and natural soils within Moscow megapolis. In: Geophysical Research Abstracts, vol. 17. EGU2015–150 (2015)
91. Vorobyova, L.A.: Soil Chemical Analysis. MSU, Moscow (1998)
92. Vasenev, V.I.: How does urbanization affect spatial variability and temporal dynamics of soil organic carbon in the Moscow region? Ph.D. thesis, Wageningen University, Wageningen, Netherlands (2015)

Modelling and Mapping Urban Soils

J.J. Stoorvogel[1(✉)], I. van Manen[1], and V.I. Vasenev[1,2]

[1] Soil Geography and Landscape Group, Wageningen University, Wageningen, The Netherlands
Jetse.Stoorvogel@wur.nl

[2] Agrarian-Technological Institute, Peoples' Friendship University of Russia, Miklukho-Maklaya Str., 6, 117198 Moscow, Russia

Abstract. Urbanization is an important trend in global land cover change and seriously impacts the soil resources. However, there is no clear definition for urban areas. As a result estimates of urbanization and its effects on soil resources vary widely. Urbanization can be modelled in different ways with a specific focus on environmental conditions, temporal dynamics and spatial patterns. The processes underlying urbanization require a hybrid approach that combines the different methods. Similarly, the unique conditions of the urban environment require specific surveying techniques for the soil resources. The global debate on urbanization and its environmental impacts calls for a rapid standardization of definitions and methodologies to come with proper information on rates and impacts.

Keywords: Soil survey · Land cover modelling · Soil classification

1 Introduction

National soil surveys often excluded the urban areas despite the fact that it was clear that soil surveys could contribute to urban planning. The Soil Survey Manual [11] recommended urban areas to be mapped intensively with 2nd order at scales between 1:12:000 and 1:31,000. However, many soil surveys aimed at agricultural development or forestry and excluded the urban areas (Fig. 1). Over the past decades the interest of the surveyors clearly changed and with a new focus on environmental impacts and carbon sequestration, the need for a better insight in soil resources is now being recognized. Soil classifications are now adapted to include the anthropogenically changed soils. The Soil Taxonomy [12] was updated on the basis of the recommendations of the International Committee on Anthropogenic Soils (ICOMANTH) to include an Anthropic Epipedon but also a range of different subgroups for human-altered and human-transported soils: the Anthraquic subgroup for irrigated rice fields, the Anthrodensic subgroup for compacted soils, the Anthropic subgroup for soils with artefacts, the (Happlo-) plaggic subgroups for soils with a plaggenepipedon, The Anthroportic subgroup for soils with human-transported material, and the Anthraltic subgroup for soils developed in human-altered topsoils. Also the World Reference Base for soil resources [6] was updated and now includes various anthropogenic diagnostic horizons like the Anthraquic and Hydragric horizons for paddy soils and the

V.I. Vasenev et al. (eds.), *Megacities 2050: Environmental Consequences of Urbanization*, Springer Geography, https://doi.org/10.1007/978-3-319-70557-6_5

Fig. 1. Urban soils as they are excluded from the 1:50,000 Dutch soil survey.

Hortic horizon for enriched topsoils but also at the highest level with the definition of Anthrosols and Technosols. However, although the soil classification has been adapted, very little attention has been paid to modelling and mapping of urban soils. This chapter aims to further explore the main issues related to modelling the expansion of urban areas and mapping soils in urban areas.

2 What are Urban Areas?

Although the definition of urban areas seems rather trivial, in practice, the definition seems to rely heavily on the application. The soil survey manual defines urban land as land that is mostly covered by streets, parking lots, buildings, and other structures ofurban areas. This definition mainly excludes areas like parks and other green areas in the urban environment and focuses mostly on the sealed areas. McIntyre *et al.* [8] provide an overview of different definitions of the area. Their review shows that different elements are considered. Urban areas can be defined on the basis of (i) the fraction of sealed areas, (ii) population density, (iii) energy use, and (iv) the function of land (*e.g.*, residential, industrial, and recreational). In addition scale plays a role, an individual building is not an urban area, but an area consisted of "houses and lawns" is considered to be an urban area [8]. Or there needs to be a minimum population (density). Although it may be obvious that different applications define urban areas in different ways, the multitude in definitions clearly hampers scientific studies and discussions on urban soils. Basically, different groups are talking about different things. If one looks at the global scale, the different definitions of urban areas can lead to tremendous differences in the urban area estimates as illustrated in Table 1. This means that one should be very careful in the selection of a particular map particularly if the area of urban areas plays an important role. For other applications this is less important. A commonly acclaimed effect of urbanization is that the expansion of urban areas takes

Table 1. The global area of urban areas according to different sources.

Source	Urban area (10^3 km^2)	% of land surface	Reference
GlobCover	477	0.22%	Bontemps *et al.* [1]
GRUMP	5,283	2.51%	Ciesin *et al.* [3]
ISA[a]	8,675	4.11%	
Population density[b]	12,676	6.01%	Ciesin [2]
Average	6,778	3.21%	

[a]Based on the global night-time lights map using the conversion provided by Elvidge *et al.* [4]
[b]Re-interpreted using the criteria provided by Short Gianotti *et al.* [10]

Table 2. The area of urban areas covering land that is potentially suitable for agriculture.

Base map	Area of fertile soils covered by urbanization (10^3 km^2)	% of urban area that covers soils potentially suitable for agriculture
GlobCover	275	61.8%
GRUMP	2,966	57.4%
ISA[a]	4,941	58.2%
Population density[b]	5,983	42.7%
Average	3,541	55.0%

[a]Based on the global night-time lights map using the conversion provided by Elvidge *et al.* [4]
[b]Re-interpreted using the criteria provided by Short Gianotti *et al.* [10]

place on fertile agricultural land and that urbanization can be considered a threat to food security. An overlay of the different maps of urban areas from Table 1 over the S-world soil map of the world [13] shows that despite the large differences in the total area, the relative impact of urbanization on land is similar (Table 2). All maps indicate that urban areas cover around 55% soils that are suitable or agricultural use.

3 Modelling Urbanization

Urbanization can be monitored closely through *e.g.,* remote sensing. However, predicting urbanization into the future is more challenging, while the evaluation of potential effects of the urbanization process on global or local soil resources is pivotal. In land cover dynamics modelling in general three different approaches are being followed relying on: (i) regression analysis, (ii) trend analysis using Markov chains, and (iii) cellular automata. Various modelling approaches can be applied. The models based on regression analysis, look for environmental conditions where certain land cover classes occur through *e.g.,* a logistics regression. A good example is the CLUE model by Verburg and Overmars [15]. In the case of urbanization one can expect that urbanization will take place on relatively flat terrain in the proximity of existing urban areas. An alternative modelling approach is a trend analysis, where the probability of a particular land cover change relies on changes in the past. A good example is provided by Muller and Middleton [9]. In the case of urbanization it can be expected that there is a certain sequence of land use change prior to the urbanization: nature – pasture – extensive agriculture – intensive agriculture – peri-urban agriculture – urban areas. A particular location in the sequence helps to identify the probability of urbanization. Finally, there are the cellular automata in which land cover changes strongly depend on the surrounding environment (*e.g.*, Fuglsang *et al.* [5]). Of course, this is very obvious in urbanization trends where urban areas expand rather than that new urban areas develop. The three modelling approaches are very distinct and base themselves on the environmental conditions, temporal trends, and spatial patterns. In reality, often more hybrid methods are implemented. Markov chains are, for example, stratified on the basis of environmental conditions, or spatial parameters (*e.g.*, distance to roads) are included in the regression models. The proper modelling of urbanization requires an integrated approach that can make use of each of the modelling approaches. However, this has repercussions for the data requirements: auxiliary environmental data are required for the regression models, time series are required for the Markov chains, and high resolution maps are required for the cellular automata. In addition, very strict definitions of urban areas are required). The latter is particular true (and problematic) for the time series.

4 Mapping Urban Soils

Standard soil surveying techniques rely on aerial photo interpretations and intensive field work with a high density of soil observations [11]. Initially soil classification systems were not suited to deal with the urban environment. This has been resolved with the recent updates of *e.g.*, the Soil Taxonomy and the World Reference Base for Soil Resources. However, we are still facing the problem that the surveying techniques are not suited for the urban environment. The specific characteristics of urban soils with abrupt changes and management effects require new approaches to be developed. Recently soil surveying techniques have received significant updates through the introduction of geostatistics and the intensive use of auxiliary and legacy data. The so-called digital soil mapping [7] has proven to be very effective in agricultural and natural environments. However, its application in urban environments is still hampered by the specific urban conditions [14]: (i) useful auxiliary data on management history and functional zones are often lacking whereas this may be one of the main soil forming factors, (ii) the abrupt changes require very high observation densities, (iii) typically, urban areas are found as islands in a landscape resulting in very clustered observations and problems to interpolate, and (iv) soil profile have a very specific build up with the so-called cultural layer within the soil profile. The specific conditions in the urban areas require specific soil surveying techniques as attempts to map them with standard soil surveying techniques or digital soil mapping are doomed to fail.

5 Conclusions

With the rapid urbanization worldwide, there is an urgent need for standardization of the definitions of urban areas, modelling of the urbanization models, and soil surveying techniques for the urban environment. Although many of the building blocks are there, the scientific community needs to put them together and come up with clear answers for the global debate.

Acknowledgements. The publication was prepared with the support of the "RUDN University program 5-100". The project was partly supported by RFBR projects Nos 15-34-70003, 15-54-53117 and 16-34-00398 and Russian Science Foundation project No 14-27-00133.

References

1. Bontemps, S., Defourny, P., van Bogaert, E.: GLOBCOVER 2009, Products description and validation report. ESA and Université Catholique de Louvain, 53 p. (2010). doi:10.1594/PANGAEA.787668
2. CIESIN: Gridded Population of the World, Version 4 (GPWv4): Population Count (2016). http://sedac.ciesin.columbia.edu/data/set/gpw-v4-population-count. Accessed 22 Sep 2016
3. CIESIN, Columbia University, IFPRI, The World Bank, CIAT: Global Rural-Urban Mapping Project, Version 1: Urban Extents Grids. SEDAC, Columbia University (2011)
4. Elvidge, C.D., Tuttle, B.T., Sutton, P.C., Baugh, K.E., Howard, A.T., Milesi, C., Nemani, R.: Global distribution and density of constructed impervious surfaces. Sensors **7**(9), 1962–1979 (2007)
5. Fuglsang, M., Münier, B., Hansen, H.S.: Modelling land-use effects of future urbanization using cellular automata: an Eastern Danish case. Environ. Model Softw. **50**, 1–11 (2013)
6. IUSS Working Group WRB: World Reference Base for Soil Resources 2014, update 2015. International soil classification system for naming soils and creating legends for soil maps. World Soil Resources Reports No. 106. FAO, Rome (2015)
7. McBratney, A.B., Mendonca Santos, M.L., Minasny, B.: On digital soil mapping. Geoderma **117**, 3–52 (2003). doi:10.1016/S0016-7061(03)00223-4
8. McIntyre, N.E., Knowles-Yánez, K., Hope, D: Urban ecology as an interdisciplinary field: differences in the use of urban between the social and natural sciences. Urban Ecosyst. **4**(1), 5–24 (2000)
9. Muller, M., Middleton, J.: A Markov model of land-use change dynamics in the Niagara region, Ontario. Can. Landscape Ecol. **9**, 151–157 (1994)
10. Short Gianotti, A.G., Getson, J.M., Hutyra, L.R., Kittredge, D.B.: Defining urban, suburban, and rural: a method to link perceptual definitions with geospatial measures of urbanization in central and eastern Massachusetts. Urban Ecosyst. **19**, 823–833 (2016)
11. Soil Survey Division Staff: Soil survey manual. Soil Conservation Service.U.S. Department of Agriculture Handbook 18. Washington DC. 315 p. (1993)
12. Soil Survey Staff: Keys to Soil Taxonomy, 12th ed. USDA-Natural Resources Conservation Service, Washington, DC. 372 p. (2014)
13. Stoorvogel, J.J., Bakkenes, M., Temme, A.J.A.M., Batjes, N.H., ten Brink, B.J.E.: S-world: a global soil map for environmental modelling. Land Degrad. Dev. **28**, 22–33 (2017)
14. Vasenev, V.I., Stoorvogel, J.J., Vasenev, I.I., Valentini, R.: How to map soil organic stocks in highly urbanized region? Geoderma **226–227**, 103–115 (2014)
15. Verburg, P.H., Overmars, K.P.: Combining top-down and bottom-up dynamics in land use modeling: exploring the future of abandoned farmlands in Europe with the Dyna-CLUE model. Landscape Ecol. **24**, 1167 (2009)

Influence of Latitudinal Zonality on Some Chemical Properties of Urban Soils

Irina A. Martynenko[1(✉)], Alexandra V. Maksimovich[1], Joulia L. Meshalkina[1,2], Jetse J. Stoorvogel[3], and Aleksey M. Yaroslavtsev[2,4]

[1] Soil Science Faculty, Moscow Lomonosov State University, 1-12 Leninskie Gory, Moscow 119991, Russia
martynenko.irina@soil.msu.ru

[2] LAMP, Russian Timiryazev State Agrarian University, 49 Timiryazevskaya st., Moscow 127550, Russia

[3] Soil Geography and Landscape Group, Wageningen University, P.O. Box 47, 6700 AA Wageningen, The Netherlands

[4] Agro-Technological Institute, RUDN University, 6, Miklukho-Maklay st., Moscow 117198, Russia

Abstract. In this study, the database of properties of urban soils was created to assess the influence of zonal features of natural factors of soils, situated in urban areas. The database contains more than 135 cities located in different natural zones all over the world from the Arctic tundra to equatorial tropics. A comparison based on two features: soil organic carbon (SOC) and water-extracted pH. A number of statistical indicators such as average value, minimum and maximum values, standard deviation, variation, correlation coefficients were calculated for each zone and feature.

The analysis of the results showed that the variation in carbon content values in urban soils is significantly higher in comparison with background soils. This fact can be explained by the influence of anthropogenic factor on the processes of organic matter reduction and accumulation in the soil. At the same time the average carbon content in urban soils in most cases is much higher than in natural soil and in general, there is a trend: the difference between SOC in urban and natural soils increases to the north and gradually decreases to the south. However, a comparative analysis of soil-bioclimatic zones neither in climatic nor in facial series there is no trend can be observed. This means that the characteristics of the SOC of urban soils are not zonal and the observed trend can be explained by the intensity of human impact. The comparison of the average values of pH in the urban and natural has not shown any statistically significant difference. However the comparison of maximum values of water-extracted pH for the groups showed a clear pattern in northern hemisphere: increasing of the alkalinity in the humid and semihumid areas from the one side, and acidification in the more southern arid and semi-arid areas from the other side, that may indicate the presence of zonal trend.

Keywords: Transformation of soil cover · Soil organic carbon · pH · Data base · Soil-ecological regions

V.I. Vasenev et al. (eds.), *Megacities 2050: Environmental Consequences of Urbanization*, Springer Geography, https://doi.org/10.1007/978-3-319-70557-6_6

1 Introduction

The Earth soil cover is an irreplaceable ecological basis of the biosphere functioning, humankind life and activity, since it is media for matter and energy complex exchange processes between the atmosphere, Earth's crust, hydrosphere and all living organisms. Now already half of the planets soil is under anthropogenic change [6, 7, 27]. One of the most important environmental problems facing our time is urbanization. Occupying slightly more than 1% of the land area, urban areas are located in practically all natural zones from sub-polar to equatorial latitudes [20]. Urban population growth in recent decades was so swift that the urban environment is no longer able to meet many of the biological and social needs of modern man [1, 26, 27].

Despite the fact that the transformation degree of an urban soil and its causes in the literature studied very well, but among a large number of publications dedicated to investigation of the soils properties in urban areas, there is virtually no research about impact level of natural areas conditions on soil properties. However, the question of interest is - if there is a trend in the, for example, pH and organic carbon distribution of the transformed urban soils. On the one hand, urbanization became the source of specific organic matter, which is unspecific for natural soils (municipal and household waste, atmospheric dust, soot, compost-based reclamation mixtures, etc.), which can lead to an increase in SOC stock. On the other hand, the natural organic substances flow is disturbed by the annual removal of plant litter, regular lawn mowing etc. [15, 24, 26]. Type of territory pollution can also affect the pH of urban soils and plant productivity in different ways [5, 15]. Thus, a high variation of soil properties is typical even for urban sites of similar land use [10, 13, 15, 17, 21]. So how close are the properties of urban soils of Murmansk (Russia), located north of the Arctic Circle and, for example, of Ibadan (Nigeria).

The aim of this study was a comparative analysis of the soil organic carbon (SOC) and the water-extracted pH of urban soils and natural (background) soil surrounding areas in different soil-bioclimatic belts and soil-ecological regions.

2 Methods

To solve this problem, the urban soil properties database was created, based on the analysis of scientific publications and our own data. The database stores soils upper horizon properties in urban areas of 138 cities around the world (Fig. 1). Cities grouping by soil-bioclimatic areas has been done in accordance with the Soil-Ecological Regions Map from Resources and Environment World Atlas (1998), scale 1:60 000 000. The urban soil properties database analysis revealed that cities are located in 14 different soil-ecological regions. Set of natural zonal soils was determined by the map FAO/UNESCO Soil Map of the World 1: 5 000 000 (2016) [8]. The SOC and pH values of background zonal soils were obtained from the Harmonized World Soil Database (HWSD) ver. 1.2 (2012). Maps were created with MapInfo Professional v. 12.5. The calculation of statistical parameters was carried out in Statistica v. 10.0.

Not only typical urban soils (Technosols) were included in the database of urban soil properties, but also slightly disturbed soils of urban forests, parks, lawns and green

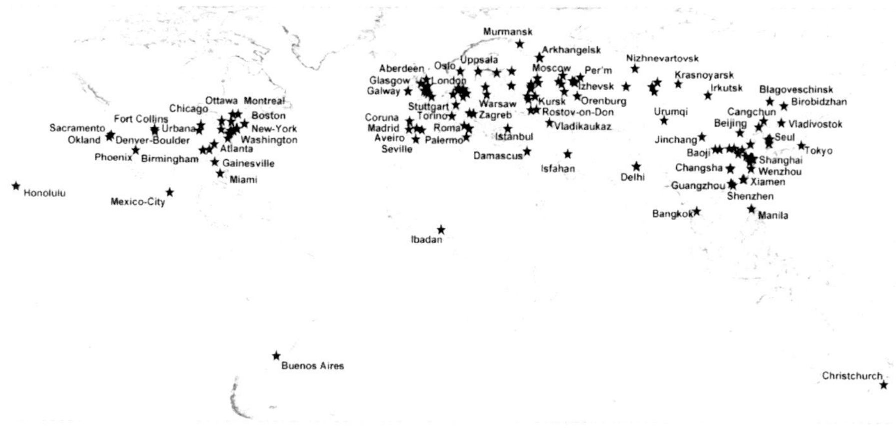

Fig. 1. Cities locations map from the urban soils database

areas intra building, located in different geomorphological positions inside the city, were included too.

Most of the cities included in the created database are situated in Russia [2, 3, 11, 30], USA [14, 19, 25], in Western European countries [4, 16, 18, 23] and in China [28, 32].

All the data, that were presented in the research were included in the database in the original units and the information of the territory land use. The base of properties was then supplemented with the GPS location of each city, soil-bioclimatic region, measurement methods and other important information, if it was mentioned in the original research. The each data has the full reference to the source.

As there were many different units, that are common in each country, the data on the carbon content needed to be converted to the general unit – SOC (% from Total C). The average value was calculated in the cases, when the several researches have been found.

To define a set of background zonal soils there was created special overlay layer between the cities and the soils of the world maps, then buffer zones of 10–15 km radius around the towns was drawn, depending on the size of the city. Soils that fall within the boundaries of the buffer zone were treated as typical. For each city there have been identified from one to three natural soils, including intrazonal alluvial and volcanic soils.

Besides that, each soil sample in urban cities within bioclimatic soil region and in each sample of natural soils, in areas identified by the buffer zones, the average values of SOC and pH were calculated. The preliminary conversion of logarithmic forms was produced in numeric to calculate the average pH, then the average value was recalculated and transformed in logarithmic.

In addition, the following statistical parameters: standard deviation, range, maximum and minimum values in the total sample and within groups, the difference between pH and SOC in the natural (background) soil and urban soils, the correlation coefficient of geographical coordinates of cities and the significance of the correlation coefficients, - were calculated for each sample in same soil – ecological region.

3 Results

The initial analysis of the literature has shown that there are virtually no complex researches that compare several urban soils and it's properties from different cities in the world. The few studies that have been found were comparing mainly heavy metal pollution, but not the natural features [9]. Most articles contained are data only for the topsoil (0–10 cm or 0–20 cm) with the indication of the territory land use and without any classification of the soil [9, 22, 28].

Cities that were selected in the database were divided into fifteen groups of soil-bioclimatic zones in accordance with the soil-ecological regions map from World Atlas Nature and Earth Resources (1998). The fifteen groups contains different number of cities: the sample size ranges from 1 to 21.

Analysis of the results shows that most of the cities is located in the Extrahumid Broad Forest Regions: a large area of the North and Central parts of Western Europe, the southern regions of the Pacific coast of Russia and North Atlantic areas of the U.S. and Canada. The second largest group are the cities of forest areas of the Humid Taiga - Forest regions. This group includes most cities of Russia. Territory of Semihumid Xerophytic Forest regions covers a large area in the South of Western Europe, and unites the cities in countries such as Spain, Portugal, Italy, Turkey. In China, area getting together the largest number of cities, was the area of Extrahumid Forest Region of the subtropical zone. Southern regions of arid, extra arid and desert regions, humid tropics and subtropics mostly weakly urbanized.

Statistical analysis showed that the SOC values range in the urban soils cities is much higher than in natural (background) soils (Table 1). This statement can be confirmed by earlier obtained data [24, 29].

This fact is a result of anthropogenic factors influence on the processes of reduction and accumulation of organic matter in the soil. In general, the difference is bigger in the more Northern regions, and gradually decreases to the South. To confirm the results obtained for the mean values, the dependence of the whole sample difference between SOC in urban and background soils from the latitude position of the city was investigated (Fig. 2).

The difference in organic matter content in urban soils compared to the background soils decreases in gradient from North to South. The difference in values are significantly correlated with latitude (correlation coefficient of 0.49 is significant).

In order to clarify, whether the regularity is a manifestation of zoning, the average values of the carbon content within the selected groups were distributed in a square matrix by the climatic zoning and the climate humidity degree in accordance with the map of soil-ecological regions legend structure (Table 2).

A comparative analysis of matrixes for urban and natural soils have not shown any dependencies. This means that the characteristics of the SOC content in urban soils is not zonal. In this way, the earlier pattern, which shown the gradient of the carbon excess in urban soils in comparison with natural, is not due to the zonal and, probably, due to socio-economic and natural-historical factors.

In the same way, pH (water) was analyzed. In comparison of urban and natural soils average pH values, statistically significant differences were not observed (Table 3).

Table 1. Means, standard deviations and ranges of SOC (% from Total C) content for urban and natural (background) soils for soil-ecological regions

Soil-bioclimatical belts	Soil-ecological regions	Sampling size	Urban soils			Background soils		
			Mean	St. dev.	Range	Mean	St. dev.	Range
Boreal belt	Humid Taiga - forest regions	20	6.77	4.31	14.8	1.58	0.82	2.79
Subboreal belt	Semihumid forest - steppe regions	7	6.70	4.26	12.86	2.01	0.40	1.35
	Extrahumid broad forest regions	21	2.46	2.93	10.76	2.26	0.96	3.79
Subtropical belt	Semihumid Xerophytic forest regions	11	2.22	3.24	9.9	1.69	0.92	3.09
	Extrahumid forest region	5	1.55	1.18	2.55	1.25	0.43	1.19
	Semiarid bush - steppe regions	4	1.71	3.60	3.13	1.30	0.92	1.82

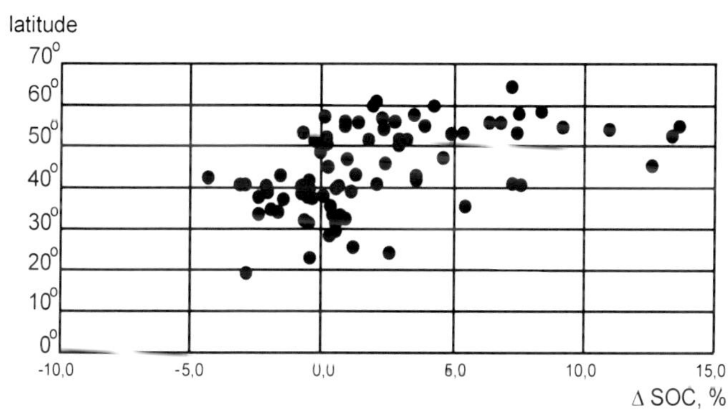

ΔSOC – the difference in carbon content between urban and background soils

Fig. 2. The dependence of the carbon content increment in urban soils from the city's latitude (Northern hemisphere)

However, a comparison of the maximum values of the urban and natural soils reveals increasing alkalinity in the following areas: humid taiga-forest, forest-steppe semiramide, extragenic forests of the subtropics and the Mediterranean region semihumid dry forests. The data of the following regions contains 2/3 of all values for urban soils. This implies that urban soils, compared to natural, in most cases have a more alkaline reaction medium.

For results verification, dependence of the difference between the pH values in urban and natural soils from the city location latitude was also studied (Fig. 3). The analysis showed that the correlation coefficient is 0.45, which is significant. Thus

Table 2. The distribution of the SOC (% from Total C) content average values for soil-ecological regions considering climate and facies

Soil-bioclimatic belts	Moisture regions					
	Extra-humid	Humid	Semi-humid	Semiarid	Arid	Extra-arid
Urban soils						
Boreal belt	No data	6.77	3.98	–	–	–
Subboreal belt	2.46	No data	6.70	5.63	No data	No data
Subtropical belt	1.55	3.04	2.22	1.71	No data	1.40
Tropical belt	No data	2.46	0.12	No data	No data	No data
Background soils						
Boreal belt	–	1.58	2.43	–	–	–
Subboreal belt	2.26	–	2.01	1.57	–	–
Subtropical belt	1.25	1.86	1.69	1.30	–	0.75
Tropical belt	–	1.72	2.98	–	–	–

Table 3. Means, standard deviations and ranges of pH (water) values for urban and natural (background) soils for soil-ecological regions

Soil-bioclimatical belts	Soil-ecological regions	Sampling size	Urban soils		Background soils	
			Mean	Range	Mean	Range
Boreal belt	Humid Taiga - forest regions	20	5.87	2.95	5.02	2.2
Subboreal belt	Semihumid forest - steppe regions	8	7.21	1.9	6.78	0.51
	Extrahumid broad forest regions	7	5.19	3.25	5.35	1.84
Subtropical belt	Semihumid Xerophytic forest regions	8	6.95	1.45	5.36	1.52
	Extrahumid forest region	4	6.85	0.48	6.08	1.54
	Semiarid bush - steppe regions	7	5.19	3.25	5.35	1.84

confirmed the dependence obtained from the analysis of maximum values of pH in regions: urban soils have a more alkaline environment than in natural conditions.

Despite the fact that the anthropogenic factor significantly affects urban soil, bringing together the values of cities soils acidity in Northern and more Southern latitudes, we can clearly trace some zonal trend (Table 4).

In humid and semihumid regions the pH value of the aqueous extract in urban soils increase from acidic to slightly alkaline range from the tropical belt to the boreal belt. In the range from extra humic to extra arid there is a clear pattern of mean values both for urban and natural soils. In the boreal zone pH decreases from humid regions to arid one. In the subtropical belt, the acidity decreases when moving from humid to arid regions. There are not enough values to estimate patterns in the tropical belt, but the same trend as in the subtropics can be observed.

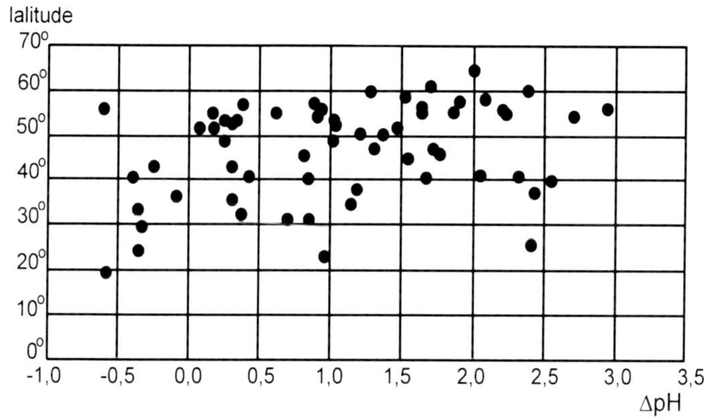

Fig. 3. The dependence of the increment of pH (aqueous extract) in urban soils from the city's latitude (Northern hemisphere).

Table 4. The distribution of the pH average values for soil-ecological regions considering climate and facies

Soil-bioclimatic belts	Moisture regions					
	Extra-humid	Humid	Semi-humid	Semiarid	Arid	Extra-arid
Urban soils						
Boreal belt	No data	5.87	7.90	–	–	–
Subboreal belt	5.19	No data	7.94	6.30	No data	No data
Subtropical belt	6.85	5.38	6.95	7.38	No data	7.46
Tropical belt	No data	4.60	5.20	No data	No data	No data
Background soils						
Boreal belt	–	5.02	4.97	–	–	–
Subboreal belt	5.35	–	7.00	6.44	–	–
Subtropical belt	6.08	5.30	5.36	6.07	–	7.48
Tropical belt	–	5.00	5.79	–	–	–

4 Conclusions

In urban soils the average carbon content is higher than in natural soils. There is a Northern gradient of the carbon excess in urban soils in comparison with natural one, however, this pattern is probably not zoned, and, apparently, has a socio-economic nature. On the one hand, natural (background) soils of boreal belt (Podzols, Albeluvisols, Cambisols) and extrahumid broad forest regions of subboreal belt (Cambisols) have a low content of SOC. On the other hand, these regions account biggest part of the historical towns and megapolises of Europe, Russia, USA and Canada, characterized by high contents of SOC.

In urban soils the average pH value is not statistically significantly different from the corresponding values in natural soils. However, while comparing the respective maximum values there is a pattern of more alkaline environment of urban soils in comparison with natural one appeared. The gradient of increasing alkalinity of urban soils in comparison with natural soils from South to North forms a trend of convergence on the level of soil acidity cities located in different climatic zones, reflecting the influence of anthropogenic factor.

Acknowledgements. This study has been performed with support by grant #14-120-14-4266-ScSh and by grant according to the Agreement № 02.A03.21.0008 of the Russian Federation Ministry of Education and Science, grant #ScSh-10347.2016.11 of the Leading Scientific Schools, and grant # 15-16-30007 of the Russian Science Foundation. We also thank Tatiana Prokof'eva for advice and manuscript review.

References

1. Agarkova, M.G.: Ecological and genetic characteristics of urban soils ecosystems (by the example of the Leninsky district of Moscow). Abstract of the Ph.D. thesis. Biol. Sci. 03.00.27. Moscow (1991). (in Russian)
2. Alexandrova, A.B.: Soil-ecological conditions of landscapes formation of Kazan. Abstract of the Ph.D. thesis. Biol. Sci. 03.00.16. Kazan (2004). (in Russian)
3. Baboshkina, S.V., Puzanov, A.V.: Chemical composition of soils and plants of various anthropogenic territories, Barnaul. world Sci. Cult. Edu. **3**, 14–18 (2008). (in Russian)
4. Beesley, L., Dicknson, N.: Carbon and trace element mobility in an urban soil amended with green waste compost. J. Soils Sediments **10**, 215–222 (2010). https://doi.org/10.1007/s11368-009-0112-y
5. Churkina, G.: Modeling the carbon cycle of urban systems. Ecol. Model. **216**, 107–113 (2008). https://doi.org/10.1016/j.ecolmodel.2008.03.006
6. Dobrovol'skiy, G.V.: Soil degradation – a threat to global economic crisis. Age Globalization **2**, 54–65 (2008)
7. Dobrovolskiy, G.V.: The place and role of soil in the biosphere and human life. Bull. Use Prot. Nat. Res. Russ. **2**, 23–27 (2011)
8. FAO/UNESCO Soil Map of the World. Scale 1:5 000 000 (2016). http://www.fao.org/soils-portal/soil-survey/en/
9. Guitao, S., Zhenlou, C., Shiyuan, X., Ju, Z., Li, W., Chunjuan, B., Jiyan, T.: Potentially toxic metal contamination of urban soils and roadside dust in Shanghai, China. Environ. Pollut. **156**, 251–260 (2008). https://doi.org/10.1016/j.envpol.2008.02.027
10. Godwin, C., Chen, G., Singh, K.K.: The impact of urban residential development patterns on forest carbon density: an integration of LiDAR, aerial photography and field mensuration. Landsc. Urban Plann. **136**, 97–109 (2015). https://doi.org/10.1016/j.landurbplan.2014.12.007
11. Gorbov, S.N.: Soil of urbolandscape of rostov-on-don, their ecological status and evaluation of pollution. Abstract of the Ph.D. thesis. Biol. Sci. 03.00.27. Rostov-on-Don (2002). (in Russian)
12. Harmonized World Soil Database (version 1.2). FAO, Rome, Italy and IIASA, Laxenburg, Austria. FAO/IIASA/ISRIC/ISS-CAS/JRC (2012). http://www.fao.org/soils-portal/soil-survey/en/

13. Bae, J., Ryu, Y.: Land use and land cover changes explain spatial and temporal variations of the soil organic carbon stocks in a constructed urban park. Landsc. Urban Plann. **136**, 57–67 (2014). https://doi.org/10.1016/j.landurbplan.2014.11.015
14. Kaye, J.P., McCulley, R.L., Burke, I.C.: Carbon fluxes, nitrogen cycling, and soil microbial communities in adjacent urban, native and agricultural ecosystems. Glob. Change Biol. **11**, 575–587 (2005). https://doi.org/10.1111/j.1365-2486.2005.00921.x
15. Lehmann, A., Stahr, K.: Nature and significance of anthropogenic urban soils. J. Soils Sediments **7**, 247–260 (2007). https://doi.org/10.1065/jss2007.06.235
16. Lorenz, K., Kaundler, E.: Biochemical characterization of urban soil profiles from Stuttgart, Germany. Soil Biol. Biochem. **37**, 1373–1385 (2005). https://doi.org/10.1016/j.soilbio.2004.12.009
17. Lorenz, K., Lal, R.: Biogeochemical C and N cycles in urban soils. Environ. Int. **35**, 1–8 (2009). https://doi.org/10.1016/j.envint.2008.05.006
18. Papa, S., Bartoli, G., Pellegrino, A., Fioretto, A.: Microbial activities and trace element contents in an urban soil. Environ. Monit. Asses. **165**, 193–203 (2010). https://doi.org/10.1007/s10661-009-0938-1
19. Pavao-Zucherman, M.A.: The nature of urban soils and their pole in ecoligical restoration in cities. Restor. Ecol. **16**, 642–649 (2008). https://doi.org/10.1111/j.1526-100X.2008.00486.x
20. Pouyat, R., Groffman, P., Yesilonis, I., Hernandez, L.: Soil carbon pools and fluxes in urban ecosystems. Environ. Pollut. **116**, 107–118 (2002). https://doi.org/10.1016/S0269-7491(01)00263-9
21. Raciti, S.M., Groffman, P.M., Jenkins, J.C., Pouyat, R.V., Fahey, T.J., Pickett, S.T.A., Cadenasso, M.L.: Accumulation of carbon and nitrogen in residential soils with different land-use histories ecosystems **14**, 287–297 (2011). https://doi.org/10.1007/s10021-010-9409-3
22. Rawlins, B.G., Vane, C.H., Kim, A.W., Tye, A.M., Kemp, S.J., Bellamy, P.H.: Methods for estimating types of soil organic carbon and their application to surveys of UK urban areas. Soil Use Manag. **24**, 47–59 (2008). https://doi.org/10.1111/j.1475-2743.2007.00132.x
23. Rodriguesa, S., Pereiraa, M.E., Duartea, A.C., Ajmone-Marsanf, F., Davidsonc, C.M., Grčmand, H., Hossackb, I., Hursthouseb, A.S., Ljungg, K., Martinif, C., Otabbongg, E., Reinosoe, R., Ruiz-Cortése, E., Urquhartc, G.J., Vrščaj, B.: Mercury in urban soils: a comparison of local spatial variability in six European cities. Sci. Total Enviro. **368**, 926–936 (2006) https://doi.org/10.1016/j.scitotenv.2006.04.008
24. Rozanova, M.S., Prokof'eva, T.V., Lysak, L.V., Rakhleeva, A.A.: Soil organic matter in the Moscow state university botanical garden on the Vorob'evy hills. Eurasian Soil Sci. **49**, 1013–1025 (2016). https://doi.org/10.1134/S106422931609012X
25. Schindelbecka, R.R., van Esa, H.M., Abawib, G.S., Wolfec, D.W., Whitlowc, T.L., Guginob, B.K., Idowua, O.J., Moebius-Clune, B.N.: Comprehensive assessment of soil quality for landscape and urban management. Landsc. Urban Plann. **88**, 73–80 (2008). https://doi.org/10.1016/j.landurbplan.2008.08.006
26. Stroganova, M., Miagkova, A., Prokofieva, T., Skvortsova, I.: Soils of Moscow and Urban Environment. PAIMS, Moscow (1998)
27. Stroganova, M.N.: Urban soils: genesis. Systematics and environmental significance (by the example of Moscow). Thesis of doctor of biological sciences. Moscow (1998)
28. Sun, Y., Ma, J., Li, C.: Content and densities of soil organic carbon in urban soil in different function districts of Kaifeng. J. Geograph. Sci. **20**, 148–156 (2010). https://doi.org/10.1007/s11442-010-0148-3
29. Vasenev, V.I., Prokof'eva, T.V., Makarov, O.A.: The development of approaches to assess the soil organic carbon pools in megapolises and small settlements. Eurasian Soil Sci. **6**, 685–696 (2013). https://doi.org/10.1134/S1064229313060100

30. Voyt'uk, E.A.: Accumulation of heavy metals in soil and plants in the urban environment (on the example of Moscow). Abstract of the Ph.D. thesis. Biol. Sci. 03.02.08. Ulan-Ude (2011). (in Russian)
31. World Atlas: Resources and Environment. I, II. Vienna: Ed.Hoelzel; IG RAS, Moscow (1998)
32. Zhang, J., Pu, L., Peng, B., Gao, Z.: The impact of urban land expansion on soil quality in rapidly urbanizing regions in China: Kunshan as a case study. Environ. Geochem. Health **33**, 125–135 (2011). https://doi.org/10.1007/s10653-010-9326-6

Heavy Metals in Urban Soils of the Yamal Region

Ivan Alekseev(✉), E.V. Abakumov, and George Shamilishvili

Saint Petersburg State University, Saint Petersburg, Russia
ivanalekseev95@gmail.com

Abstract. Soil chemical properties are essential for the functioning of soils in the polar biome. This study aimed to study the concentrations of heavy metals in urban soils of Harsaim, Aksarka, Salekhard, Harp and Labytnangi. At 12 sites 23 soil samples were collected at depths of 0–5 cm and 5–20 cm. Heavy metals were detected with X-ray fluorescent analyzer "Spectroscan-MAX". The values obtained were compared with the Approxible Permissible Concentrations and Maximum Allowable Concentrations adopted in Russia. The study of soil samples from different settlements let to reveal the characteristic features of soil contamination of individual settlements with heavy metals to compare them with each other. The vast majority of samples are characterized by excess of Maximum Allowable Concentrations for arsenic, which should indicate a high regional background of this element. For a more adequate assessment of the levels of total soil contamination (Saet's index) Zc during its calculation it was used not only average arithmetic values of coefficient of concentration (Kk), but also its average geometric values. Most of the soil samples are characterized by non-hazardous levels of total soil contamination. The study showed a statistically significant difference in content of heavy metals for the 0–20 cm layer of the soils for three elements (Cu, Zn, Ni).

Keywords: Heavy metals · Soil contamination · Yamal peninsula

1 Introduction

The functioning of the polar biome strongly relies on its soils. The geochemical regime directly connects soils with their function as agent for accumulation, migration and transformation. In addition, soils play very significant role for various ecosystem services. Trace metals on the one hand are naturally present in parent rock and soils occurring in the form of sulfides, oxides, silicates, and carbonates [2].The aim of this study is to asses the contamination levels of the urban soils of central part of Yamal autonomous region.

2 Methodology

In August 2015, the diversity in the geochemical conditions of soils in the Yamal-Nenets Autonomous District was studied. One of the directions of this work was connected with study of urban soils of the Yamal-Nenents autonomous District

V.I. Vasenev et al. (eds.), *Megacities 2050: Environmental Consequences of Urbanization*, Springer Geography, https://doi.org/10.1007/978-3-319-70557-6_7

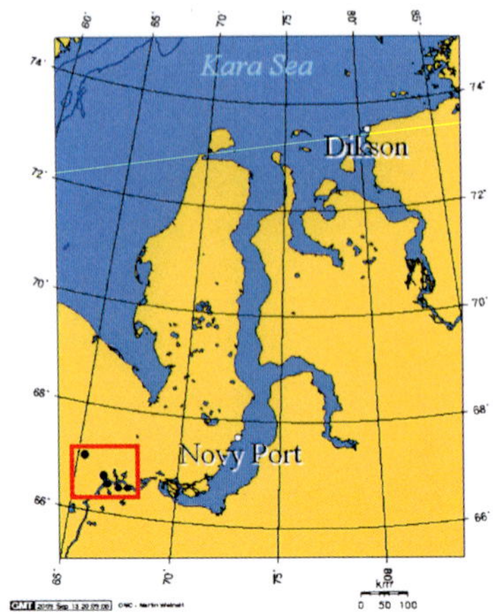

Fig. 1. The area of study (within Yamal autonomous region)

settlements. Soils in Harsaim, Aksarka, Salekhard, Kharp and Labytnangi were studied (Fig. 1).

During field work 12 sites were studied and samples were taken from a depth of 0–5 cm and 5–20 cm (a total of 23 samples). Soil diagnostics were performed in accordance with the classification and diagnostics of Russian soils [9]. Soil samples have been collected in industrial (Labytnangi, Kharp), residential (Salekhard) and recreational functional zones (Aksarka, Kharsaim). Laboratory analysis was conducted in the Komi Scientific Centre Laboratory of the Russian Academy of Sciences. Heavy metals contents (Pb, Cd, Cu, Ni, Zn, As and Hg) were determined with an X-ray fluorescent analyzer "Spectroscan-MAX" [7]. The values obtained were compared with the permissible concentrations and maximum allowable concentrations adopted in Russia named in GN 2.1.7.2511-09 [4], GN 2.1.7.2041-06 [5] and SanPin 42-128-4433-87 [8]. Cd's regional soil background concentrations of the Yamal peninsula [6], Hg's Earth crust clarke [3] and concentrations of the rest trace elements in natural sandy soil from the Beliy island [1] were used in calculations.

Statistical analysis of the data was obtained in the environment of the analytical software interface STATISTICA 10. The choice of the method with Kruskal-Wallis ANOVA criterion was connected with small amount of analyzed samples. These samples in addition are different in sense of distribution laws. Kruskal-Wallis criterion is non-parameterized alternative for one-dimensional dispersive analysis.

3 Results and Discussion

The study of soil samples from different settlements reveals the characteristic features of soil contamination of individual settlements with heavy metals (Table 1). The vast majority exceeded the maximum allowable concentrations for arsenic, which should indicate a high regional background of this element. For a more adequate assessment of the levels of total soil contamination Zc during its calculation not only average

Table 1. Heavy metals contents in urban soils of Yamal region

Horizon (depth, cm)	Cu, mg/kg	Pb, mg/kg	Cd, mg/kg	Zn, mg/kg	Ni, mg/kg	As, mg/kg	Hg, mg/kg
Kharsaim							
O (0–5)	10.5	36.0	0.39	143	13	2.6	0.04
G (5–20)	8.0	7.0	0.19	27	16	3.7	0.01
O (0–5)	1.6	2.3	0.10	5.7	4.0	0.8	0.01
Gtur (5–20)	1.6	2.3	0.10	5.4	4.4	0.7	0.01
Aksarka							
O (0–5)	3.8	4.0	0.14	11.6	8.3	2.1	0.005
G (5–20)	9.1	7.6	0.17	25	15	2.9	0.014
O (0–5)	5.5	10.0	0.20	17	12	2.4	0.009
G (5–20)	6.0	150	0.22	17	14	2.4	0.008
Salekhard							
W (0–5)	7.1	8.9	0.25	27	9.0	2.9	0.031
C (5–20)	7.5	8.9	0.22	25	9.0	3.0	0.037
W (0–5)	6.3	6.2	0.27	20	10	3.0	0.012
C (5–20)	7.0	6.7	0.22	24	11	3.4	0.012
Kharp							
W (0–5)	73	7.9	0.24	56	50	2.7	0.027
C (5–20)	80	3.0	0.1	46	19	1.0	0.009
W (0–5)	74	4.2	0.1	49	28	1.4	0.006
C (5–20)	86	3.9	0.1	49	30	1.1	0.008
C (5–20)	27	9.9	0.43	45	380	5.7	0.031
Labytnangi							
W (0–5)	9.0	9.1	0.26	31	13	2.8	0.052
C (5–20)	6.0	6.1	0.14	19	10	3.1	0.013
W (0–5)	6.2	6.6	0.14	19	9.0	2.4	0.011
C (5–20)	7.4	7.0	0.13	21	12	2.7	0.011
W (0–5)	9.5	7.3	0.16	27	17	3.1	0.017
C (5–20)	11.2	7.7	0.23	28	17	4.0	0.014
Allowable concentrations (for sandy soils)	–	32	–	–	–	2	2.100
Permissible concentrations	33	–	0.5	55	20	–	–

arithmetic values of coefficient of concentration (Kk) were used, but also its average geometric values. Most of the soil samples are characterized by non-hazardous ($Zc < 16$) levels of total soil contamination. Calculation of soil pollution index showed that the most of soil samples have values less than 16. It characterizes the soils as unpolluted.

Table 2 shows the statistically significant differences in the content of heavy metals and hydrocarbons for the 0–20 cm layer of the soils for three elements (Cu, Zn, Ni).

Table 2. Mean values of heavy metals contents for 0–20 cm layer in soil samples from urban areas of Yamal region

	Kharsaim (n = 4) mean ± SD	Aksarka (n = 4) mean ± SD	Salek-hard (n = 4) mean ± SD	Kharp (n = 5) mean ± SD	Labyt-nangi (n = 6) mean ± SD	*p* Kruskal-Wallis ANOVA	*p* One-way ANOVA
Cu	5.42 ± 4.53	6.10 ± 2.21	6.97 ± 0.50	68.00 ± 23.50	8.22 ± 2.04	0.01	<0.001
Pb	4.65 ± 2.71	7.90 ± 2.83[e]	7.67 ± 1.43	55.78 ± 2.97	7.30 ± 1.04	0.62	0.36
Cd	0.19 ± 0.14	0.18 ± 0.03	0.18 ± 0.02	00.19 ± 0.14	0.18 ± 0.05	0.48	0.87
Zn	16.27 ± 12.38[e]	17.65 ± 5.52	24.00 ± 2.94	49.00± 4.30	24.16 ± 5.15	0.03	<0.001
Ni	9.35 ± 6.07	12.32 ± 2.95	9.75 ± 0.96	31.40 ± 11.35[e]	13.00 ± 3.41	0.01	<0.001
As	1.95 ± 1.46	2.45 ± 0.33	3.07 ± 0.22	2.38 ± 1.97	3.02 ± 0.55	0.20	0.57
Hg	0.01 ± 0.01	0.01 ± 0.004	0.02 ± 0.01	0.02 ± 0.01	0.02 ± 0.02	0.20	0.60

The main peculiarity of the studied soils is the heightened contents of arsenic (As). It testifies the heightened regional background concentration of this heavy metal. Soils of Kharp settlement are characterized by excess of permissible concentrations in copper (Cu) and nickel (Ni) contents in most of soil samples. Excesses in permissible concentrations and maximum allowable concentrations in other settlements are characterized by sporadic distribution and are caused more likely with local properties of soil pollution in specific sampling places.

4 Conclusions

Urban soils of the central part of Yamal region were investigated with special reference to assessment of heavy metals concentration in topsoil. The main feature for trace elements in studied soils is heightened content of arsenic (As) in most of the soils. It could be connected with high levels of background concentration for Yamal region. Statistical analysis on heavy metals content in the upper 20 cm showed statistically significant differences for three elements (Cu, Zn, Ni). This investigation was carried out in urban areas. That is why there was a limitation of depth in sense of collecting the soil samples (0–20 cm). However, it should be noticed that profile approach is more preferable for detailed investigation of trace elements is soils. It lets to deduce the dynamics of trace elements profile distribution and take into account cryoturbation factor, which is one of the most important in pedogenesis of permafrost-affected landscapes. Data obtained could be used in further environmental researches and environmental management purposes in Yamal region key oil and gas exploration region.

Acknowledgements. This work was supported by Russian foundation for Basic research, project No 16-34-60010 and Russian President grant for young doctors of science MD-3615.2015.4.

References

1. Abakumov, E.V., Lodygin, E.D., Gabov, D.A., Tomashunas, V.M.: Soderzhanie politsiklicheskikh aromaticheskikh uglevodorodov v pochvakh Antarktidy na primere rossiyskikh polyarnykh stantsiy. Gig. I Sanitariya **1,** 30–4 (2014)
2. Antcibor, I., Eschenbach, A., Zubrzycki, A., Kutzbach, L., Bolshiyanov, D., Pfeiffer, E.-M.: Trace metal distribution in pristine permafrost-affected soils of the Lena River delta and its hinterland, northern Siberia, Russia. Biogeosciences **11**, 1–15 (2014). doi:10.5194/bg-11-1-2014
3. Greenwood, N.N., Earnshaw, A.: Chemistry of the Elements, 2nd edn. Butterworth-Heinemann, Oxford (1997). 1359 p
4. GN 2.1.7.2511-09. Approximate allowable concentrations of chemical substances in soils
5. GN 2.1.7.2041-06. Maximum allowable concentrations of chemical substances in soils
6. Moskovchenko, V.D.: Geochemistry of northern landscapes of West-Siberian Plain: structural and functional organization of the geosystems matter and ecodiagnostics. Extended Abstract of Doctoral (Geogr.) Dissertation, St. Petersburg (2010)

7. OST 10-259-2000. Pochvy. Rentgenofluorescentnoe opredelenie valovogo soderzhania tyazhelyh metallov. (in Russian)
8. SanPiN 42-128-4433-87. Sanitary norms for available concentrations of chemical compounds in soils
9. Shishov, L.L., et al.: Classification and diagnostics of Russian soils. Oikumena, Smolensk (2004)

Ecogeochemical Assessment of Soil Cover of the Ufa City, Bashkortostan

R. Suleymanov[1], R. Urazgildin[1], K. Amineva[1], A. Suleymanov[2], and E.V. Abakumov[3](✉)

[1] Ufa Institute of Biology, Ufa, Russia
[2] Moscow State University of Geodesy and Cartography, Moscow, Russia
[3] Saint-Petersburg State University, Saint-Petersburg, Russia
E_abakumov@mail.ru

Abstract. Soils of Ufa city were previously under-investigated in terms of chemical pollution and anthropogenic impacts. This study shows the different trends of soil chemical transformation under different functional load (industrial zone, parks and natural environments). Anthropogenic transformed Umbric Luvisols and Albeluvisols were studied in terms of basic chemical properties and trace elements content. Data show that oil pollution results in an accumulation of carbon and nitrogen. However, this accumulation does not correspond to real humus accumulation, but to the accumulation of black-carbon-like fractions. Redistribution of trace elements over depth is typical for the contaminated soils. Zinc concentrations do not exceed the permissible values, but the trends of zinc migration down the soil profile are evident. Lead concentrations increase in contaminated soils, but permissible concentrations are not exceeded. Urban soil contamination leads not only to the accumulation of contaminants in the upper soil horizons, but also to the redistribution of some trace elements in the soil profile.

Keywords: Soils · Urban ecosystems · Contamination · Forest-steppe zone

1 Introduction

Soils play a key role in urban ecosystems functioning. Nevertheless, soils of cities of Russian steppe region are under-investigated in terms of morphology and chemical composition. Urban soil's formation in steppe semiarid environments differs from those in humid boreal ecosystems. Here, humus enriched soil with neutral or alkaline reaction of fine earth are affected by urbic processes. Therefore, accumulation of organic carbon and changing of soil pH are less evident in urban soils. Besides, the behavior of trace elements in neutral or alkaline environments is specific. This investigation aimed to study the soils transformation in one on the biggest Russian south-east cities- Ufa.

Ufa City is a large industrial center with numerous oil processing and transportation facilities, chemical industry and a developed web of traffic routes. Ufa is the capital of huge and key transitional region – Bashkortostan, which is situated on the border of Europe and Asia. This city is a good example of urban marcolandscape at the boundary of forest-steppe and steppe zones. Due to intensive industrialization, the population of

V.I. Vasenev et al. (eds.), *Megacities 2050: Environmental Consequences of Urbanization*, Springer Geography, https://doi.org/10.1007/978-3-319-70557-6_8

the city faces various environmental problems connected with ecological risks. A comprehesive ecogeochemical assessment in the Ufa City region is lacking. An objective of this investigation is to conduct the geochemical characteristics of soils in urban forests, sanitary forest protective zones and number of parks located in different city parts.

2 Materials and Methods

Ufa locates close to the South Ural mountain ridges, on the territory of Pribelskaya upland, namely on the Belsko-Ufimskiy watershed. Downtown and the main residential areas locate on Ufa peninsula, which presents the up landed plateau, surrounded by rivers and lowlands. The city covers and extent of 708 km^2 populate by approximately 1.1 million. Climate is continental with warm summers and severe winters. The territory of the city was conventionally divided into three parts:

1. territory of high contamination (north part of the city, sanitary protective zone of petroleum factory, soil profile № 1),
2. territory of middle contamination intensity (central part of the city, park zones with natural origin of the wood stand, soil profile № 2), and
3. control plot zones (south sector of the city, natural landscape and forest, soil pit № 3).

The soils are classified as Umbric Luvisols and Albeluvisols of the forest-steppe zone. These soils have some evidences of anthropogenic transformation, especially in case of soil pits 1 and 2. Soil pits are sub-layered by clay textured bedrocks debris. The forest stands include oak, birch, pine, spruce, maple, poplar and larch. The soil profiles can be characterized by their morphological formulas:

- Profile 1: O(0-2)-AU(2-25)-AB(25-39)-BI(39-60)-D(60-80),
- Profile 2: O(0-3)-AU(3-29)-AUeB(29-49)-BT(49-72)-C(72-90), and
- Profile 3: AU(0-24)-AB(24-37)-BI (37-49)-C(49-110).

The soil profiles were samples and analyzed for soil organic carbon, extractable forms of nitrogen, total and extractable forms of the phosphorous, exchangeable cations and pH using the techniques by Arinushkina [1], trace elements concentrations (Cu, Cd, Zn, Fe, Pb) were determined by atomic absorption [2] and hydrocarbons content were assessed by gravimetry [3].

3 Results and Discussion

The three soil profiles were classified as Umbric Albeluvisols with weak features of podzolisation and a distinct line of the carbonates accumulation. This outcome corresponds well to the previous works [4, 5]. The humus content is estimated as medium (Table 1). In the control soil profile humus content is 4.08 in upper layer and 0.68% in parent material fine earth. In the soils with low contamination load these values are 3.90 and 0.89% correspondingly. Soils with the highest contamination load also have the highest humus content in the uppermost layer (5.19%). The values of extractable

Table 1. Soil chemical characteristics

Horizon, depth, cm	pH in water	Humus, %	Extractable nitrogen, mg/kg	Poshourous		Ca^{2+}	Mg^{2+}	Petroleum hydrocarbons mg/kg
				Bulk, mg/kg	CmolP/kg	CmolP/kg		
Plot 1 (strong contamination load)								
$A_1$0–25	6.74	5.19	364	141.4	3.9	57	13	580
A_1B 25–39	7.26	3.14	112	118.9	5.3	51	10	470
B39–60	7.46	1.45	84	106.6	4.5	44	10	80
C61–90	7.58	1.00	68	98.4	3.2	37	9	32
Plot 2 (medium contamination load)								
$A_1$0–29	6.25	3.90	196	135.2	5.4	42	13	280
A_1B 29–45	6.56	2.88	154	101.5	5.10	49	10	150
B45–72	7.48	1.35	70	106.6	5.3	47	10	trace
C72–100	7.61	0.89	56	100.4	4.8	41	10	0
Plot 3 (control soil)								
$A_1$0–24	6.18	4.08	294	100.4	4.1	34	11	0
A_1B 24–37	5.68	2.13	126	147.5	4.0	34	10	0
B37–49	5.67	1.24	70	86.1	5.5	34	13	0
C49–100	7.22	0.68	56	100.4	4.6	40	13	0

nitrogen forms are high in all the investigated soils, but the increased concentrations of nitrogen were revealed in the soil with highest contamination load. Soil of strong and medium contamination zones have increased concentrations of the petroleum products, whereas these substances were not revealed in the soil of control plot.

High correlation rate was revealed for humus and petroleum products content (r = 0.93) and for nitrogen content and petroleum products values (r = 0.82), thus accumulation of petroleum products in soil leads to apparent increment of the carbon and nitrogen. This is caused by the activities of the oil industry in these locations. Phosphorous concentrations are low: 4.0 mg/100 g in control soils and 3.9 mg/kg in contaminated soil. The same is found for the bulk phosphorus concentration: 10.4 mg/kg in control and 141.4 mg/kg in strongly contaminated soil. Soil acidity shows a trend to decrease from control soil to low and strong contaminated ones. This is caused by increment of exchangeable Ca and Mg cations concentrations in soils at the same sequence.

Thus, data obtained show that oil contamination lead to increasing of carbon and nitrogen contents in soils. Similar results were found earlier [6, 7], while it was shown that the increment of carbon content was connected with accumulation of insoluble remnants or pseudo-humus in soils. This substance has an origin that is related to black carbon. It is therefore concluded that the accumulation of carbon as a result of anthropogenic impact is not real, but apparent.

The trace elements concentrations prove that copper concentration is low and the lowest values are typical for the most contaminated soil. Maximal accumulations of the copper refers to the upper soil horizons and inside these upper layers there is some redistribution of the copper with increasing of concentration on the depth about 30–50 cm and further decreasing of concentrations. Iron accumulation in upper horizons of contaminates soil is well expressed, permissible concentrations are strongly exceeded. Redistribution of the iron concentration within the depth also typical for contaminated soils investigated (Table 2).

Zinc concentration does not exceed the permissible values, but the profile curves of its distribution also show the trends of zinc migration download the soil profile. The lead concentrations increases in contaminated soil, but does not show the trends of permissible concentrations exceeding. The maximal concentration of lead referred to the uppermost soil horizons. Similar results were found for Orenburg city [8].

Table 2. Heavy metal contents in soils of Ufa City.

Depth, cm	Cu	Cd	Zn	Fe	Pb
	mg/kg				
Plot 1 (strong contamination load)					
0–10	10.09	0.48	48.14	16334	20.67
10–20	10.96	0.45	43.73	16470	22.5
20–30	10.58	0.44	44.53	19118	20.7
30–40	8.50	0.25	40.06	16466	13.53
40–50	5.07	0.24	25.79	13213	14.96
50–60	6.35	0.24	20.16	11543	13.98
60–70	6.77	0.04	9.44	10166	10.1
70–80	4.56	0.04	11.43	6441	9.69
80–90	2.87	0.04	1.17	5831	2.37
Plot 2 (medium contamination load)					
0–10	11.83	0.2	45.49	26435	12.15
10–20	10.49	0.02	46.72	18586	11.07
20–30	15.42	0.15	45.19	25759	9.72
30–40	12.06	0.14	50.63	32877	7.82
40–50	9.86	0.14	48.32	34124	7.12
50–60	9.56	0.09	56.84	33681	4.85
60–70	9.76	0.1	51.2	29865	5.96
70–80	10.12	0.08	41.65	30035	4.07
80–90	9.16	0.08	36.31	26729	4.47
90–100	7.54	0.08	38.7	26710	4.49
Plot 3 (control soil)					
0–10	12.98	0.15	40.14	15284	7.70
10–20	12.85	0.13	27.40	18458	2.63
20–30	12.70	0.15	41.72	22729	7.67
30–40	13.17	0.14	55.16	30113	7.43
40–50	14.65	0.15	56.34	45860	7.68
50–60	16.58	0.14	68.52	45558	8.57
60–70	13.46	0.13	54.60	37766	6.73
70–80	9.36	0.10	33.10	25916	6.45
80–90	8.68	0.09	31.20	23579	6.62
90–100	9.85	0.11	35.04	26100	5.98

4 Conclusion

First attempt to assess the soil profile distribution of trace element conducted for Ufa city. Soil from landscapes with different functional load investigated. An analysis of trace elements concentrations shows low copper concentration and that the lowest values coincide with the most strong contaminated soil. Maximum accumulations of copper takes place in the upper soil horizons. Inside these upper layers there is some redistribution of copper with an increased concentration at a depth of 30–50 cm. Iron accumulation is well expressed in the upper horizons, where permissible concentrations are strongly exceeded. Also some redistribution of zinc was revealed in studies soils. Data obtained shows that contamination leads not only to the accumulation of the main contaminant in the upper soil horizons, but also to the redistribution of some trace elements in the soil profile. It should be taken into account, while one elaborates new requirements to the maximal permissible concentration rates. The problem is that the official requirement to assess the contaminant content only refers to the uppermost horizons of top soil. This results in an underestimation of the real rates of contaminant accumulation in the soils. In conditions of forest-steppe urban environment redistribution of trace elements result in underestimation of real levels of pedo-environment contamination.

Acknowledgement. This work was supported by Russian Scientific Foundation, project No. 17-16-01030 "Soil biota dynamics in chronoseries of posttechnogenic landscapes: analyses of soil-ecological effectiveness of ecosystems restoration".

References

1. Arinushkina, E.V.: Manual on the chemical analysis of soils. Moscow State University (1970)
2. Brishke, M.E.: Atomic absorption spectrochemistry analyses (Methods of analythical chemistry). 224 p. M. Chemistry (1982)
3. Mukatanov, A.H.: Soil-ecological regions of Bashkortostan. Trans. Bashkirian Acad. Sci. **1** (2), 62–69 (1996)
4. Federal environmental regulations PND F 16.1.41–04. Quantitative chemical analysis of soils. In: Methods of Measurement of Mass Concentration of Petroleum Products in the Soil Samples by Gravimetric Method, Moscow (2004)
5. Forest of Bashkotrostan. Ed. Hayretdinov A.F.: Ufa. 400 p. (2004)
6. Codina, L.A.: Geochemical diagnostics of oil contamination of soil. In: Recovery of Oil-Contaminated Soil Ecosystems, pp. 112–122. Nauka (1988)
7. Kahatkina, M.I., Tsutsaeva, V.V., Novak, A.V.: The composition of humus floodplain soils contaminated with oil. In: Rational Use of Soil and Soil Cover of Western Siberia, TSU, Tomsk, pp. 89–97 (1986)
8. Voevodina, T.S., Rusanov, A.M., Suleymanov, R.R.: Assessment of phytotoxicity and level of chemical contamination of soils of Orenburg roadside area. Vestn. Bashkir State Agrarian Univ. **3**(35), 9–13 (2015)

Lead in New York City Soils

Ireyena Li[1], Zhongqi Cheng[1(✉)], Anna Paltseva[1,2], Tatiana Morin[1], Brianne Smith[1], and Richard Shaw[3]

[1] Department of Earth and Environmental Sciences, Brooklyn College, The City University of New York, 2900 Bedford Avenue, Brooklyn, NY 11210, USA
zcheng@brooklyn.cuny.edu
[2] PhD Program in Earth and Environmental Sciences, Graduate Center of The City University of New York, 365 5th Avenue, New York, NY, USA
[3] USDA-NRCS Soil Survey, Somerset, NJ 08873, USA

Abstract. Urban soil is a sink for anthropogenic lead (Pb) and the latter is a persistent threat to human health, especially to children and the gardening population. In the past decade, several organizations have tested soil samples for Pb in New York City. Here we summarize the available soil Pb data for New York City and create a spatial distribution map. The highest Pb levels were present in the oldest parts of the city, and mostly industrial and high traffic areas. There is overlap between high Pb areas with areas of high population density and high poverty rates. The analyses help delineate parts of the city that are most affected, possible sources of Pb, and where to prioritize resources for mitigation and remediation.

Keywords: Lead · Soil · New York City · GIS map

1 Introduction

Lead (Pb) is a toxic element with well-known adverse health effects. Exposure to Pb can hinder neurological development in young children, cause hypertension in adults, premature births, elevated blood pressure and kidney disease [19, 37]. New Yorkers are more likely to have a higher Blood Lead Level (BLL) than the majority of the United States population. The geometric mean concentration of NYC adult BLL is equal to 1.79 μg/dl. Meanwhile the national average estimate is equal to 1.56 μg/dl [19]. Moreover, the NYC Department of Health and Mental Hygiene (DOHMH) found that 4.8% of 1,811 NYC adults tested had a BLL greater than 5 μg/dl [24].

Special precautions should be taken for children, who have been identified as the highest risk group [38] and are more likely to experience negative health-related effects due to elevated BLLs [20]. A national study concluded that about 900,000 American children younger than 6 years old have a BLL greater or equal to 10 μg/dl, and that many of these children reside in inner city high poverty areas. For example, approximately 69% of children in Boston (6 months to 5 years old) have a BLL greater than 10 μg/dl [37]. Although there has not been a similar widespread study conducted in NYC, the NYC Department of Health (NYCDOH) found that 7,657 children (6 months

V.I. Vasenev et al. (eds.), *Megacities 2050: Environmental Consequences of Urbanization*, Springer Geography, https://doi.org/10.1007/978-3-319-70557-6_9

to less than 6 years old) have a BLL greater than or equal to 10 μg/dl [25]. Beginning in 2012, in light of new research of even low Pb level's toxicity effect on children, Center for Disease Control and Prevention (CDC) set the reference value for lead poisoning to be 5 μg/dl [6, 14, 21]. The CDC also set a recommendation for community blood testing when a child's BLL is greater than 10 μg/dl. However, there is no safe level of Pb because even low concentrations pose a health concern [37].

Soil could be an important source of Pb exposure for many people, especially those living in urban areas. Children who spend time in playgrounds and backyards with contaminated soils are at risk for ingestion and inhalation of soil Pb. Several studies have demonstrated clear correlation between soil Pb and BLL [20, 37, 39]. A study focusing on the effectiveness of soil Pb remediation on BLL found that after soil Pb abatement, the Boston children experienced an average BLL decline of 2.44 μg/dl [37]. Moreover, a study in Detroit, Michigan found that children's BLL were higher during the summer in the months of July, August and September, than in January [39]. The author links the BLL to higher atmospheric Pb due to more suspended contaminated soil dust during the summer months, which increases the likelihood of inhaling contaminated particles.

It has been recognized that history in NYC Pb exposure most likely stemmed from leaded gasoline, lead-based paint, lead dust and airborne pollution, waste incineration, coal and oil combustion [2, 21]. Further, soil Pb may be transported into interior areas of the house and accumulate in dust by various carriers, such as, soil Pb transport by humans, pets, and wind. Although the use of leaded-gasoline and leaded-paint are banned, Pb has been deposited in the topsoil and is still a persistent source of Pb to the environment. One previous study found higher Pb levels in soil that were adjacent to roads–indicating deposition from leaded gasoline [33]. In this same study, Pb soil levels between roads and homes ranged between 20–1,060 mg/kg. Meanwhile, the homes' Pb dust levels ranged from 130–11,760 mg/kg – indicating soil contribution to Pb dust in homes. Additionally, Brooklyn has a higher median concentration of interior Pb dust than the other boroughs [5]. Brooklyn's median measured at 730 $\mu g/ft^2$, where 85% of interior Pb dust originated from soil Pb and sources other than lead-based paint.

Although soil Pb distribution in many large cities have been conducted [17], there has not yet been an extensive study for NYC. A previous study found that soil Pb in 54 NYC community gardens frequently exceeded the guidance values [21]. As gardening activities in NYC increase, the risk of ingestion or inhalation exposure of contaminants increases [9, 18, 21, 29]. In fact, inhalation of fine soil particles may prove just as harmful as ingestion [7]. An earlier study concludes that finer soil particles, smaller than 125 μm, have double the level of Pb found in larger particles, smaller than 180 μm [10].

This study will examine the spatial distribution of Pb levels in NYC and its relationship to land use, population and poverty. Soil Pb levels at different depths will also be compared between naturally deposited and fill (human transported) soils. Previous papers have emphasized the need for a large-scale Geographic Information Systems (GIS) map for a better health-based assessment in NYC [11]. Based on data collected, this paper will include a color-coded GIS map to help residents make informed decisions related to urban agriculture and to minimize health risks. This is also important for understanding health risks related to direct soil ingestion and inhalation in parks and children's playgrounds. The analyses can further identify affected parts of the city as well as where interventions and remediation resources should be prioritized.

2 Methods

2.1 Data Compilation

This research compiled Pb measurements from 2,292 soil samples. These include mostly samples submitted to the Soil Testing Lab at Brooklyn College. The garden soil heavy metal data was reported in Cheng et al. [10]. Additional samples from parks, tree pits and yards from the New York City Housing Authority were analyzed in this study using a portable XRF environmental analyzer (Innov-X, Delta Classic) [3]. The results from a Con Edison study [1], data from the National Cooperative Soil Survey [22] (unpublished), and some data from the SoilSHOP event held in Snug Harbor in Staten Island in 2015 are also included here. These are all the soil Pb data that are available for this study.

2.2 Historical Research

Historical research for this study focused on identifying old manufacturing/industrial areas with potential Pb-containing products. Several elevated soil Pb locations identified in this paper were in the same location as manufacturing/industrial areas. Descriptions of some of these manufacturing/industrial areas include the specific use of Pb-based products. The Metropolitan Transit Authority provided Environmental Impact Statements (EIS), which identified several contaminated sites. This paper compiled the contaminated sites from the EIS with the historical reproduced manufacturing/industrial maps from the New York Public Library (NYPL).

2.3 Statistical Analysis

PostgreSQL PgAdmin III, Microsoft Office Excel, IBM SPSS, and ArcGIS were used for statistical analysis. Each sample had an accompanying identification number, soil Pb level, sampling depth, land use and addresses/street names or latitude/longitudinal coordinates or GPS degree decimals. The locations with multiple samples only include the median Pb values.

This research compiled 244 samples with depth data from the NCSS [22]. Other sources used in this study did not provide corresponding data on depth and "fill" or "naturally deposited" soil origination. The average depths of samples with a range of $\pm$1–63 cm are included in this study. The 244 samples consist of human-transported (fill) soils and naturally deposited soils. The classification depends on the park sample location. Different parks have artifacts that distinguish if a sample was naturally deposited soil or was fill soil [31]. For example, fly ash would distinguish a sample as fill soil because fly ash originates from the burning of coal in boilers [16, 30, 31]. Other artifacts which distinguish a sample as fill include glass, metal, paper, plasterboard, plastic, potsherd, rubber, and treated and untreated wood [31].

Spearman Rank Correlation was calculated between population and median soil Pb level in each zip code. Population data came from the 2010 United States Census Bureau [35]. ArcGIS 10.3.1 was used to calculate the population within each zip code. Zip codes with less than three sample locations were not included. Out of 178 zip codes

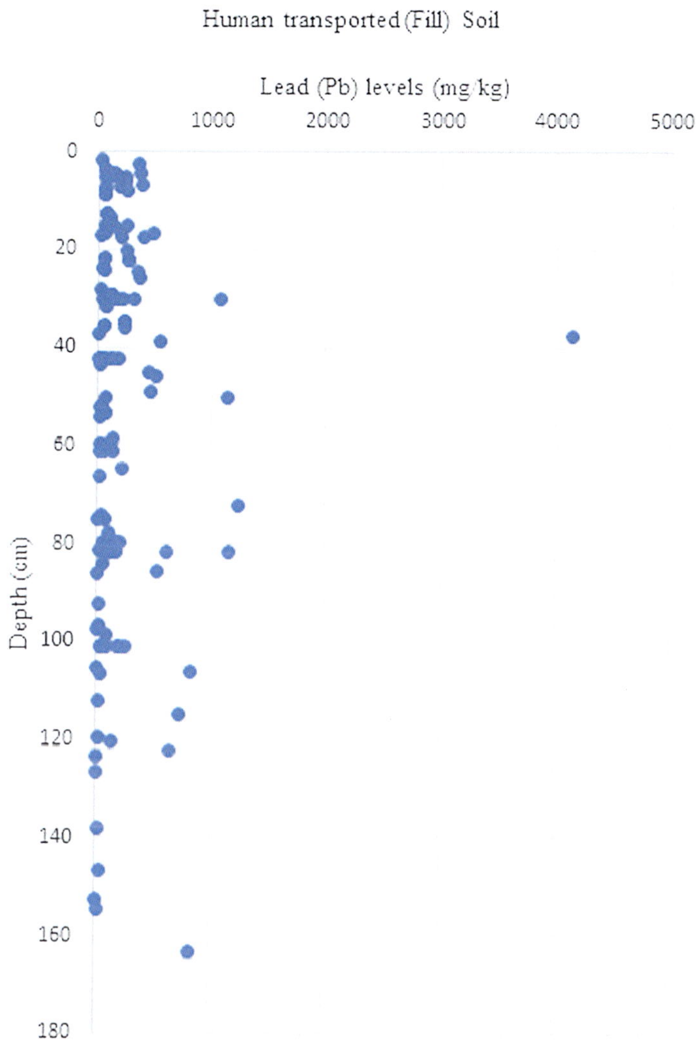

Fig. 1. A scatter plot contrasting the level of Pb vs. depth in human-transported soils. Human transported soils have little correlation between depth and Pb levels. Sixteen different locations from 157 samples are included.

in NYC, only 104 zip codes had enough samples to be included in this statistical analysis. This study utilized Spearman's Rank Correlation because it provides a non-parametric correlation of the relationship between ranked variables. Spearman's Rank Correlation is better suited for the dataset than other correlations (e.g., Pearson's correlation coefficient) because it takes into account a non-linear relationship and does not need to be specific to a normal distribution curve. It also makes no assumptions on the distributional relationship of population and soil Pb.

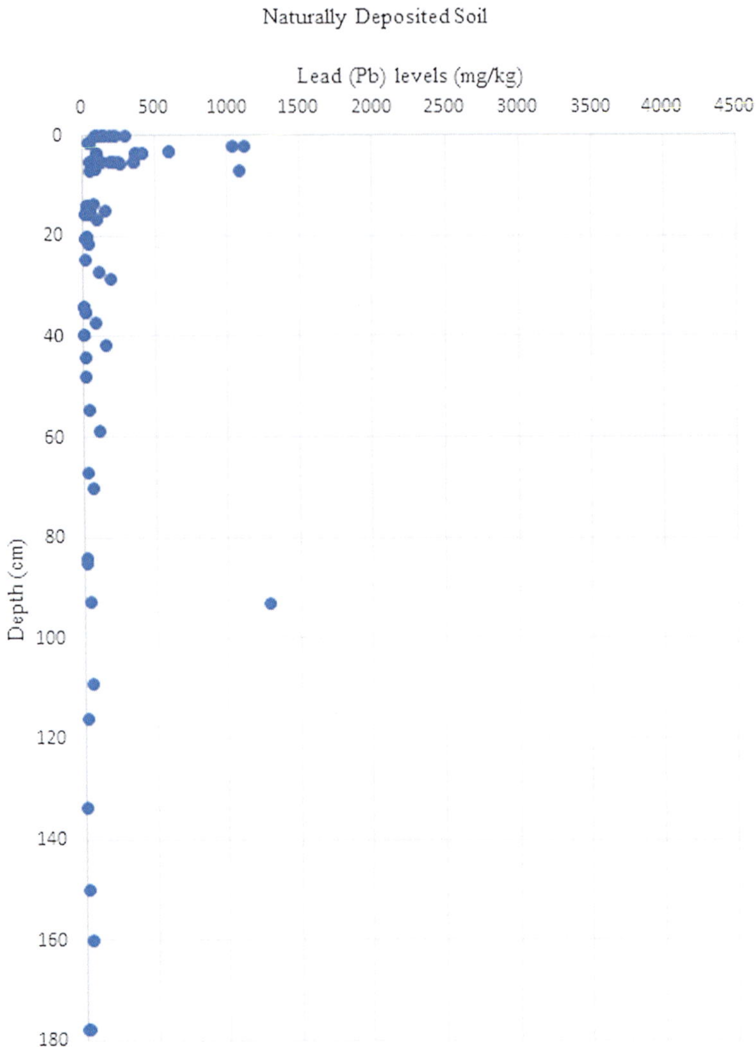

Fig. 2. A scatter plot of Pb levels vs. depth in naturally deposited soils. Naturally deposited soils have decreasing Pb levels with increasing depth levels. Fourteen different locations from 87 samples are included.

2.4 Geospatial Analysis

Use of ESRI ArcGIS and Google Earth created a visual portrayal of results for geospatial analysis. This research excludes samples with apparent errors if manual corrections proved unsuccessful. Each individual location's median Pb value calculated each zip code's median. This study combined the zip code median layer with the ZCTAS zip code boundaries [34], which were from the 2010 United States Census.

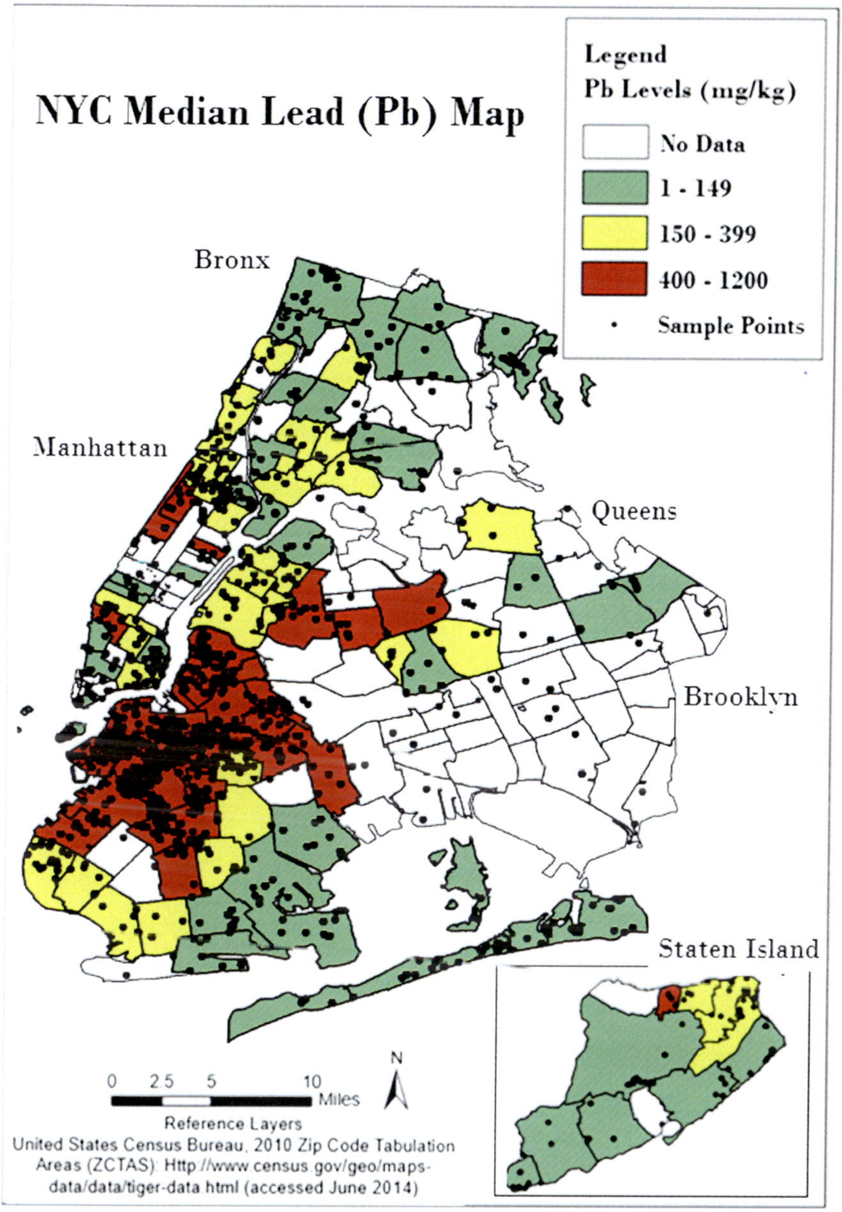

Fig. 3. The New York City (NYC) median lead (Pb) map shows soil Pb distribution in relation to zip codes. The map shows 1,230 locations from a total of 2,292 samples compiled from various sources: ESAC, DPR, Con Edison, Soil Kitchen (EPA), NCSS, and NYCHA. The highest Pb levels are centered near downtown Brooklyn, western parts of Queens, parts of downtown Manhattan, Harlem, and a small proportion of northern Staten Island.

This study includes four Pb categories that reflect different Pb thresholds. The 1200 mg/kg threshold reflects EPA standard of 1200 mg/kg for non-children play areas [36]. The 400 mg/kg threshold reflects EPA standard of 400 mg/kg for children play areas as well as the NYSDEC guidance value of 400 mg/kg [27]. The category of 1–149 mg/kg is reflective to the Center for Disease Control and Prevention (CDC) reference value of 5 μg/dl. Based on research done by the Toxics Cleanup Program Policy and Technical Support Unit, 2010, a level of 150 mg/kg of lead in soil can lead to an approximate blood lead level of 5 μg/dl.

The soil Pb map is compared to a land use map. The Land Cover Raster Data map from New York City Parks and Recreation is a background layer [26], overlaid with soil Pb levels that are expressed in circles. The location of circles is in reference to the ZCTAS zip code boundaries layer from the 2010 United States Census Bureau [34]. Larger circles represent a higher Pb level, and gradually decrease in size for lower Pb levels. ArcGIS identified 'hotspots' in large manufacturing/industrial use areas throughout NYC. This was done by using the Optimized Hotspot tool, which uses Getis-OrdGi* statistics, along with the Fishnet polygon computation option. Hotspots in this research are clusters of Pb levels greater than or equal to 400 mg/kg. ArcGIS was used to create raster files from a 1922 original map named 'Industrial Map New York City Manufacturing Industries' by the Industrial Bureau of The Merchants' Association of New York, reproduced by The New York Public Library in 2015. Spatial Statistics calculated hotspots based on the manufacturing/industrial use areas overlay and the Pb data compiled from this study. The hotspots focused on areas where Pb levels are greater than or equal to 400 mg/kg with a 90–99% confidence level.

Using the Land Cover Raster Data from the New York City Parks and Recreation [26], there is a relationship between the percentage of industrial/commercial areas (e.g., paved surfaces) and Pb levels. Calculations of a radius of approximately 20 feet and 60 feet, around each sampling point determined the percentage of paved surfaces near the soil sample. The radius creation and computation used the Focal Statistics and Extraction of Multi Values to Points in the Spatial Analyst extension tool of ArcGIS.

Data for roads with an Annual Average Daily Traffic (AADT) of greater than 30,000 vehicles was taken from the Traffic Monitoring Unit in the Highway Data Services Bureau at the New York State Department of Transportation, (1999–2013) [28]. This data overlaid a median Pb zip code map generated by this research, using a reference zip code map from the United States Census Bureau, 2010. Only roads averaging greater than 30,000 vehicles and intersecting zip codes with values within Brooklyn and Queens were included. Railroad contaminated sites [8] were also included.

3 Results

3.1 Spatial Distribution of Pb

Shown in Fig. 3 is the spatial distribution of soil Pb from the compiled datasets. On the map, each zip code is color-coded based on the median Pb levels for all the samples within the zip code area. The highest range (400–1200 mg/kg) shown in red, is present

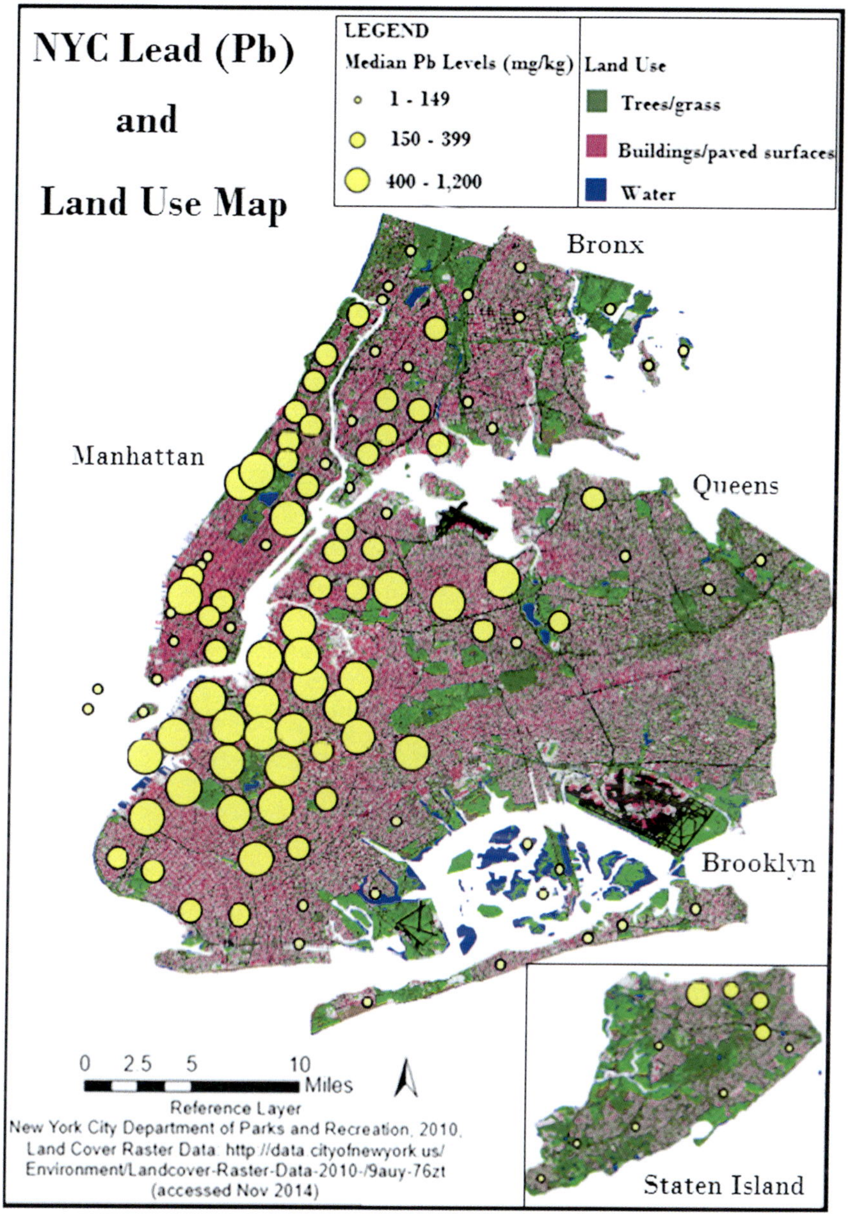

Fig. 4. Median Pb (mg/kg) zip code levels are expressed as yellow circles. The data overlays a land cover raster map from the NYC Department of Parks and Recreation, 2010, Land Cover Raster Data. Pb levels are present in the industrialized (pink) areas of the map. Industrialized areas include buildings, streets, and sidewalks. Pb levels dissipate where trees/grass (green) become dominant.

throughout various parts of northern Brooklyn (i.e. Downtown Brooklyn), near Long Island City, and in downtown Manhattan, Harlem, and a limited portion of northern Staten Island. The middle range Pb category, in yellow (150–399 mg/kg) is along the edges of the red zones, surrounding downtown Brooklyn and Long Island City. The middle category is also present along northern Staten Island, downtown Manhattan, northern Manhattan, and the South Bronx. The lowest range (1–149 mg/kg) is shown in green and is present throughout Staten Island, in isolated areas in Manhattan, and in scattered areas in the Bronx, eastern parts of Queens, the Rockaways, and southeastern parts of Brooklyn. Each black dot on the map is a sampling point. Each black dot can represent multiple samples that are from the same street address.

3.2 Correlations of Pb Levels to Land Use

Table 1 shows the percentage of samples within the four categories of Pb levels: 0–149 mg/kg, 150–399 mg/kg, 400–1,200 mg/kg, and greater than 1,200 mg/kg. Each of these categories is broken down into the following sources: gardens, parks, and public housing. The majority, 74% of samples from parks and 63% of samples from public housing are less than 150 mg/kg. In comparison, only 29% of the samples from gardens are below 150 mg/kg. It has been shown that home gardens are much more contaminated than community gardens [10]. A total of 14% of the garden samples and 2% of the parks samples contained over 1200 mg/kg of lead.

Figure 4 demonstrates a correlation between Pb levels and land use. High levels of soil Pb visually correspond with industrialized areas. For example, areas near Downtown Brooklyn and the southwestern-most part of Queens (Long Island City) have elevated soil Pb levels. Lead levels are generally lower in areas where trees/grass coverage become dominant. Overall, soil Pb levels in vegetated areas are much lower when compared to industrialized areas.

Figure 5 shows a present day Pb level map overlaying manufacturing/industrial areas in downtown Brooklyn in the year 1922. Lead hotspots (gridded boxes) that are equal to or greater than 400 mg/kg at a 90–99% confidence level are present in downtown Brooklyn. The majority of the Pb sample points are within close proximity to manufacturing/industrial sites. It is important to look at sampling hotspots because residents will have a higher incentive to submit a sample if they have a reason to believe their soil is contaminated. Similarly, Fig. 6 shows a present day Pb level map overlaying the 1922 map showing manufacturing/industrial areas along with Pb/sample

Table 1. Number of soil samples and percentages of soil Pb are shown in 4 categories: 0–149, 150–399, 400–1200 and >1200 mg/kg (ppm), for each source. This table includes 2,292 total samples from each of the sources: ESAC, NYCParks, NYCHA, EPA, and Con Edison. Large percentages of samples exceed 150 mg/kg, and 400 mg/kg.

Sources	No. of samples	Pb Level (mg/kg)			
		0–149	150–399	400–1200	>1200
Gardens	1646	29%	24%	34%	14%
Parks	117	74%	19%	5%	2%
PublicHousing	246	63%	31%	6%	0%

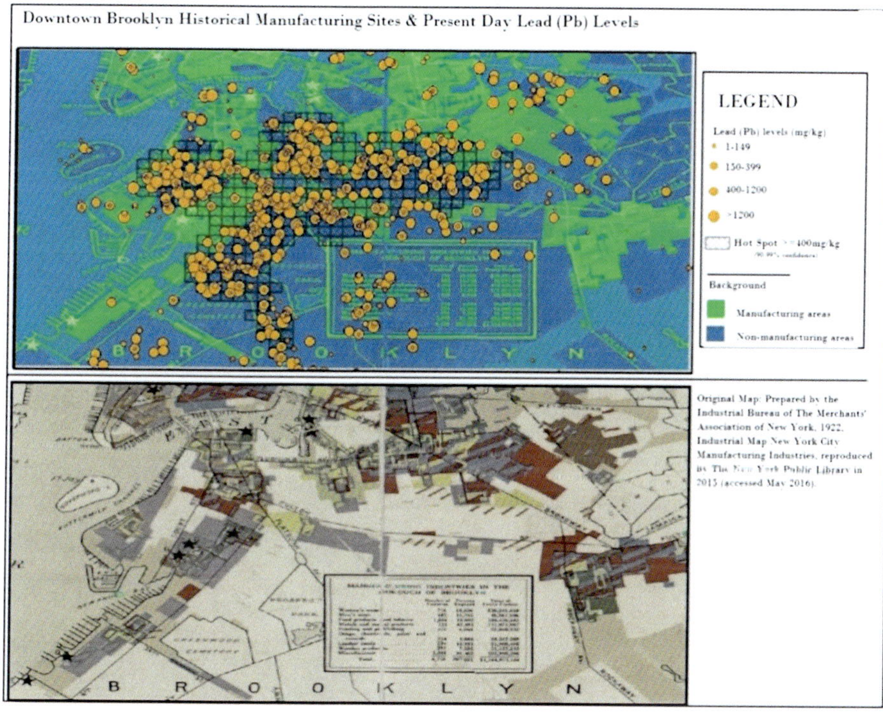

Fig. 5. Lead (Pb) levels overlaying 1922 manufacturing sites in area near Long Island City. Present day Pb hot spots (≥400 mg/kg) are clustered in 1922 manufacturing/industrial areas. This Pb hotspot is also a sample hotspot because residents are more likely to submit a sample if they suspect contamination. Indeed, these samples result in elevated levels. Therefore, residents' suspicion is often an indicator of where there are elevated Pb levels.

hotspots equal to or greater than 400 mg/kg at a 90–99% confidence level near Long Island City. Pb hotspots in Figs. 5 and 6 overlay a map originally prepared by the Industrial Bureau of The Merchants' Association of New York, 1922, (Industrial Map New York City Manufacturing Industries, and reproduced by The New York Public Library in 2015). The hot spots in Fig. 5 cluster near manufacturing/industrial areas. Within the gridded hot spot areas, 1,649 samples are present. Specific hot spot neighborhoods include Red Hook, Brooklyn Heights, Gowanus, Park Slope, Boerum Hill, Fort Greene, Williamsburg, and Bedford-Stuyvesant.

The hot spot area for Fig. 6 includes 271 samples. Similar to Fig. 5, the hot spots in Fig. 6 are near manufacturing/industrial areas. The Pb sampling hotspot in Fig. 6 is in the neighborhood called Greenpoint. Similar to Fig. 5, the most contaminated soils have a high level of resident concern and are more likely to be tested, thus creating a sampling hotspot. Based on a long history of industrial/manufacturing activity, Long Island City is likely to be a Pb hot spot for Pb in addition to Greenpoint. These analyses demonstrate the fact that due to industrial/manufacturing and railroad deposition of Pb in previous years (i.e. 1922), the Pb levels is still present today.

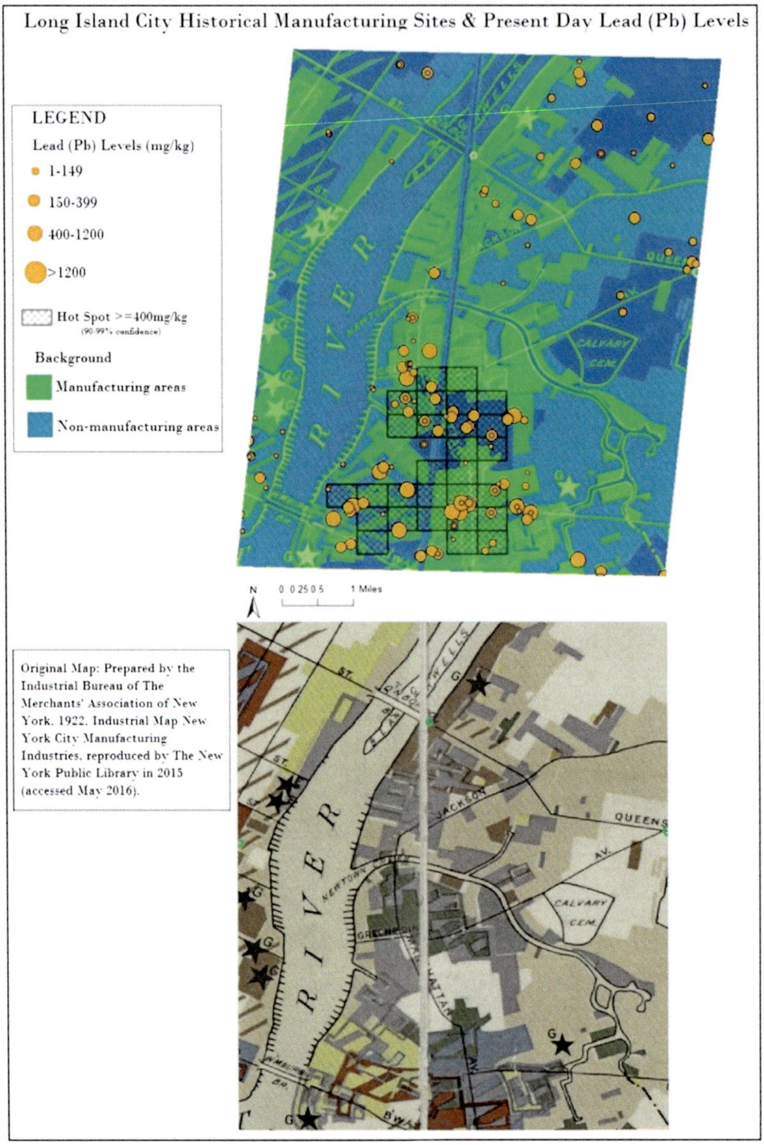

Fig. 6. Lead (Pb) levels overlaying 1922 manufacturing sites in area near Long Island City. Present day Pb hot spots (≥ 400 mg/kg) are clustered in 1922 manufacturing/industrial areas. This Pb hotspot is also a sample hotspot because residents are more likely to submit a sample if they suspect contamination. Indeed, these samples result in elevated levels. Therefore, residents' suspicion is often an indicator of where there are elevated Pb levels.

The map of Traffic (Fig. 7) portrays roads in Brooklyn and Queens with an Annual Average Daily Traffic greater than 30,000 vehicles (Traffic Monitoring Unit, 1999–2013) [28]. These data overlay a median Pb (mg/kg) zip code map. Leaded gasoline

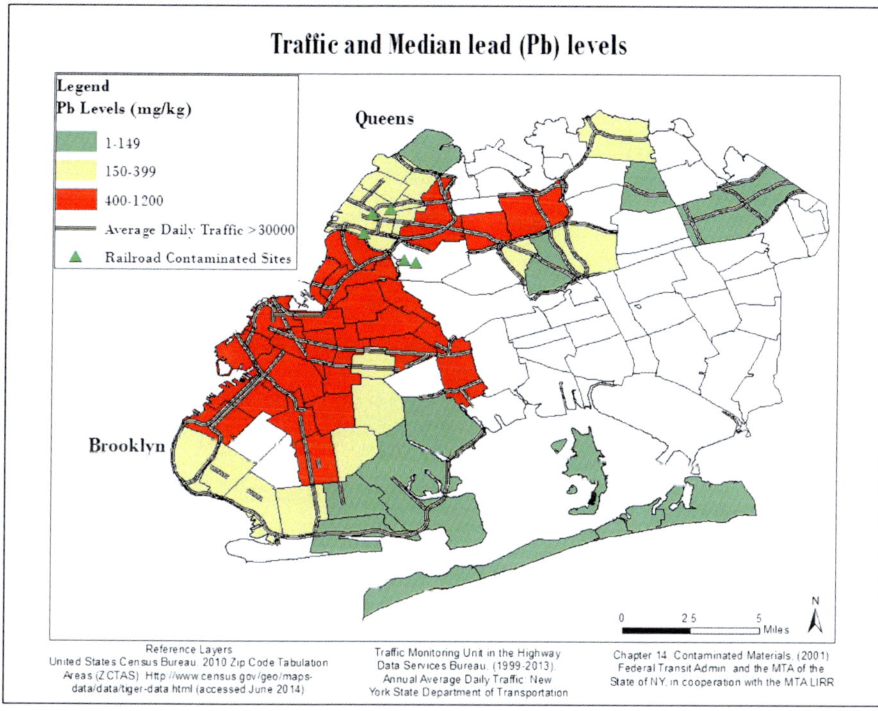

Fig. 7. Roads with Annual Average Daily Traffic >30,000 vehicles (Traffic Monitoring Unit, 1999–2013) overlay a median Pb (mg/kg) zip code map (United States Census Bureau, 2010) Vehicles on high traffic roads before the 1970s, deposited large amounts of Pb, which is reflective in higher zip code median Pb values (i.e. areas near Downtown Brooklyn and Long Island City, Queens). Railroad contaminated sites (described in Chapter 14: Contaminated Materials – East Side Access FEIS (2001), the Federal Transit Administration and the MTA of the State of NY, in cooperation with the MTA LIRR) are identified in the area near Long Island City, Queens.

prior to the 1970s deposited large amounts of Pb especially around high traffic roads. Areas around high traffic roads are consistent with zip codes with a higher median Pb value (i.e. areas near downtown Brooklyn and Long Island City, Queens). More than 50% of the 238 roads shown in Fig. 7 intersect the red zip code zones (400–1,200 mg/kg). This shows that the red zip code zones have a higher density of high traffic roads. Additionally, historical research of the Long Island City area identified four contaminated sites with elevated Pb levels coincident with historic railroad use [15].

3.3 Correlation of Pb Levels to Population Density

The Spearman Rank Correlation suggests a slight correlation between population and median Pb levels. The results of the Spearman Rank Correlation are as follows: the coefficient is equal to 0.297, the two-tailed p-value is equal to 0.003, and the number of variables is 104. Since the Spearman Rank Correlation for this study is equal to 0.297, the data has a slight correlation between population and soil Pb. The two-tailed p-value provides the likeliness of a random sample of data having the same or higher correlation coefficient, as well as the statistical significance of the Spearman Rank Correlation value.

3.4 Depth Distribution of Pb in Naturally Deposited and Human Transported Soils

The scatter plot in Figs 1 and 2 contrasts the level of Pb vs. depth in human-transported (often called "fill") soils and naturally deposited soils. Fill soils are sampled approximately between 0 to 160 cm below the surface. A total of 157 samples from 16 different locations are included. The Pb levels in fill soils range from 0 to 4,130 mg/kg, and are elevated when compared to naturally deposited soils. Fill soils have little correlation between depth and Pb levels, as high Pb levels are present at all depths. Even at a depth of 160 cm, some samples have greater than 500 mg/kgPb. Figure 2 shows a scatter plot of Pb levels vs. depth for naturally deposited soils. Naturally deposited soil samples range from the surface to as deep as 180 cm. Fourteen different locations and 87 samples are included. Lead levels decrease with depth for naturally deposited soil samples. Nearly all of the elevated Pb levels are present at the surface or very close to the surface (approximately less than 10 cm), where Pb levels can be greater than 1000 mg/kg. One sample at an approximate depth of 90 cm has 1,000 mg/kg of Pb – it is not clear whether this may be due to a mis-classification of this pedon (likely should be fill soil). When sample depth is greater than 10 cm, maximum Pb levels decrease to less than 250 mg/kg.

4 Discussion

4.1 Potential Sources of Pb

This study has found that high levels of Pb dominate throughout various parts of Brooklyn and Queens (Fig. 3), and these areas overlap with old industrialized areas (Figs 5 and 6) and high traffic roads (Fig. 7). Historically, leaded gasoline, lead-based paint, and many other lead-based products were widely used up until the phase-out period around 1980's and 1990's [13]. The use of leaded gasoline and lead-based paint resulted in massive amount of Pb deposition into the soil. Manufacturing activities involving lead-bearing products could also have contaminated soil with large amounts of Pb. All these sources could have contributed to the Pb hotspots identified in Downtown Brooklyn and areas near Long Island City, Queens (Figs. 5 and 6). In particular, an estimated four to five million metric tons of Pb from car exhaust had been released into the environment from 1929 to 1986 throughout the United States [32]. In previous research, the center of Brooklyn and the southwestern-most part of Queens (Long Island City) were high traffic density hotspots [5]. Figure 7 shows numerous high traffic roads through Brooklyn and Queens that annually average greater than 30,000 vehicles on a daily basis.

Lastly, soild waste incineration might also have played a role in soil deposition especially during the 20th century [10]. Deposition of 34 million tons of refuse incineration throughout NYC landfills caused the release of 1 million tons of air pollutants. The deposition of these air pollutants settles onto the topsoil.

4.2 Population, Poverty and Pb Levels

In areas with larger populations, soil Pb is often elevated. Based on the Spearman Rank Correlation, a slightly positive relationship is present between population and Pb levels. Brooklyn has continued to be the borough with the highest population in both 1980 (shortly after the Pb phase-out) and 2010 (the most recent census), and has the highest soil Pb level [4, 10]. Unfortunately, more populated areas seem to have higher levels of Pb, disproportionally exposing a larger number of individuals to Pb health risks.

In addition, buildings that are likely to have lead-based paint are often in neighborhoods with a larger population and high poverty, due to the presence of old housing. Comparing results in Fig. 3 to citywide poverty data [23] shows that elevated Pb levels are more prevalent in areas with high poverty levels. A history of some of the highest poverty percentages dominate in the South Bronx, northern Manhattan, the area south of the Brooklyn/Queens border, southern and central parts of Brooklyn and the eastern half of the Rockaways. Although some neighborhood poverty levels are decreasing, the higher levels of Pb correspond with the areas' history of high poverty during which Pb product use prevailed. Likewise, in year 2000, 35% of children who lived in a home with an average income of equal to or less than $30,000 had lead-based paint hazards. In comparison, only 19% of all other housing units (with incomes greater than $30,000) had a lead-based hazard [17]. There is a lack of funding for mitigating and education on Pb risks. Remediation for Pb deposition in these areas was inadequate, an environmental justice issue that have not received enough attention. Resources should be prioritized to mitigate the Pb hazards in this economically disadvantaged population.

4.3 Limitations of the Dataset

The Pb data from various sources are not evenly distributed spatially throughout NYC. Examples of areas that are not included but are likely to have high Pb levels are near the John F. Kennedy International Airport, the LaGuardia Airport, and in areas with a history of high vehicle traffic and high industrial and manufacturing use. It is also important to note that the median Pb levels in Fig. 3 could have a bias towards elevated Pb levels. This is because garden owners have an increased incentive to submit soil samples if they suspect their soil is contaminated. In Fig. 4, Pb levels are lower in areas where trees/grass areas become dominant. Although Pb is lower in less industrialized settings (Fig. 4), a sampling bias may be present because trees/grass areas are not of major concern and sample collection is likely to be less frequent.

For the different data sources compiled in this paper, sampling protocols vary. The majority of garden samples are composite samples from 3–10 different locations within a site. Gardeners were instructed to collected samples from surface to 6 inches, or 0–2 inches depending on purpose of soil testing. However, since sampling was conducted by garden owners, it is unknown if sampling protocols were always closely followed. Some sources in this paper did not have a specified sampling protocol, they most likely followed the EPA sampling standard protocol, which includes both composite and spot sampling [13]. The EPA spot sampling includes removing the top 2 inches of soil, and using only the top ½-inch as a sample, whereas the composite sampling protocol is similar to the protocol use for garden soil sampling for the Soils Lab at Brooklyn College.

5 Conclusion

This study produced the first comprehensive soil Pb map in NYC, identified contaminated parts of the city and the spatial relationship of Pb levels to land use, roads, poverty, population, soil depth, and soil type. High Pb levels closely relate to industrialized land use and the oldest parts of the city. Industrialized land use includes historical Pb deposition from manufacturing/industrial facilities and major roads with heavy traffic. Data were not available for many areas in NYC, therefore future sampling should be done to provide a more thorough understanding of Pb distribution. Although most Pb products had been banned since 1970's in the United States [12], Pb is not biodegradable and is still present in topsoil in almost all urban areas. Unless removed, Pb presents a health risk to the urban population, especially to children and to the gardening population, in the foreseeable future. High Pb areas often coincide with areas of high population and high poverty, leading to a disproportionally high percentage of people exposing to Pb hazards and those without resources for reducing their health risk from Pb. Resources should be prioritized to these areas in the short term.

Acknowledgments. Colleen Simon and Zaw Win Naing from Midwood High School assisted in data entry and testing of some of the samples with a handheld XRF analyzer. The authors appreciate the following organizations for allowing us to include their soil Pb in this study: Brooklyn College Soil Testing Lab, New York City Department of Parks and Recreation (DPR), and the New York City Housing Authority (NYCHA), USDA National Cooperative Soil Survey, US EPA SoilSHOP (formerly Soil Kitchen).

References

1. Azzolina, N.A., Kreitinger, J.P., Skorobogatov, Y., Shaw, R.K.: Background concentrations of PAHs and metals in surface and subsurface soils collected throughout Manhattan, New York. Environ. Forensics **17**(4), 294–310 (2016)
2. Biasioli, M., Barberis, R., Ajmone-Marsan, F.: The influence of a large city on some soil properties and metals content. Sci. Total Environ. **356**, 154–164 (2006)
3. Billets, S.: Innovative Technology verification report XRF technologies for measuring trace elements in soil and sediment XCALIBUR ELVAX XRF analyzer. U.S. Environmental Protection Agency, Washington, DC, EPA/540/R-06/006 (NTIS PB2006-109028) (2006)
4. Bureau of Census: 1980 Census of Population: United States Department of Commerce (1983)

5. Caravanos, J., Weiss, A.L., Blaise, M.J., Jaeger, R.J.: A survey of spatially distributed exterior dust loadings in New York City. Environ. Res. **100**, 165–172 (2006)
6. Center for Disease Control and Prevention (CDC): Low level lead exposure harms children: a renewed call for primary prevention. Advisory Committee on Childhood Lead Poisoning Prevention of the CDC (2012)
7. Chaney, R.L., Mielke, H.W., Sterrett, S.B.: Speciation, mobility, and bioavailability of soil lead. In: Davies, B.E., Wixson, B.G., (eds.) Science Reviews, pp. 105–109 (1988)
8. Federal Transit Administration and the Metropolitan Transit Authority of the state of New York, in cooperation with the MTA Long Island Railroad (2001). Chapter 14: Contaminated Materials. East Side Access – Final Environmental Impact Statement. 14-1–14-26. Accessed 20 May 2016
9. Cheng, Z., Lee, L., Dayan, S., Grinshtein, M., Shaw, R.: Speciation of heavy metals in garden soils: evidences from selective and sequential chemical leaching. J. Soils Sed. **11**, 628–638 (2011)
10. Cheng, Z., Paltseva, A., Li, I., Morin, T., Huot, H., Egendorf, S., Su, Z., Yolanda, R., Singh, K., Lee, L., Grinshtein, M., Liu, Y., Green, K., Wai, W., Wazed, B., Shaw, R.: Trace metal contamination in New York City garden soils. Soil Sci. **180**, 1–8 (2015)
11. Datko-Williams, L., Wilkie, A., Richmind-Bryant, J.: Analysis of U.S. soil lead (pb) studies from 1970 to 2012. Sci. Total Environ. 854–863 (2014)
12. Environmental Protection Agency (EPA): EPA Takes Final Step in Phaseout of Leaded Gasoline. EPA press release (1996)
13. Environmental Protection Agency (EPA): 40 CFR Part 745 Lead: Requirements for Lead-Based Paint Activities in Target Housing and Child-Occupied Facilities. Final Rule. Federal Registrar. 45,777–45,825 (1996)
14. Henry, H., Naujokas, M.F., Attanayake, C., Basta, N.T., Cheng, Z., Hettiarachchi, G.M., Maddaloni, M., Schadt, C., Scheckel, K.G.: Bioavailability-based in situ remediation to meet future lead (Pb) standards in urban soils and gardens. Environ. Sci. Technol. **49**, 8948–8958 (2015)
15. Industrial Bureau of The Merchants' Association of New York. Industrial Map New York City Manufacturing Industries, reproduced by The New York Public Library in 2015 (1922)
16. Kapička, A., Petrovský, E., Ustjak, S., Macháčková, K.: Proxy mapping of fly-ash pollution of soils around a coal-burning power plant: a case study in the Czech Republic. J. Geochem. Explor. **66**, 291–297 (1999)
17. Levin, R., Brown, M., Kashtock, M., Jacobs, D., Whelan, E., Rodman, J., Schock, M., Padilla, A., Sinks, T.: Lead exposures in U.S. children, 2008: implications for prevention. Environ. Health Perspect. **116**, 1285–1293 (2008)
18. McBride, M., Shayler, H., Spliethoff, H., Mitchell, R., Marquez-Bravo, L., Ferenz, G., Russel-Anelli, J., Casey, L., Bachman, S.: Concentrations of lead, cadmium and barium in urban garden-grown vegetables: the impact of soil variables. Environ. Pollut. **194**, 254–261 (2014)
19. McKelvey, W., Gwynn, R., Jeffery, N., Kass, D., Thorpe, L., Garg, R., Palmer, C., Parsons, P.: A biomonitoring of lead, cadmium, and mercury in the blood of New York City adults. Environ. Health Perspect. **115**, 1435–1441 (2007)
20. Mielke, H., Gonzales, C., Smith, M., Mielke, P.: The Urban Environment and children's health: soils as an integrator of lead, zinc, and cadmium in New Orleans, Louisiana, U.S.A. Environ. Res. **81**, 117–129 (1999)

21. Mitchell, R., Spliethoff, H., Ribaudo, L., Lopp, D., Shayler, H., Marquez-Bravo, L., Lambert, V., Ferenz, G., Russel-Anelli, J., Stone, E., McBride, M.: Lead (pb) and other metals in New York City community garden soils: factors influencing contaminant distributions. Environ. Pollut. **187**, 162–169 (2014)
22. National Cooperative Soil Survey, National Cooperative Soil Characterization Database. http://ncsslabdatamart.sc.egov.usda.gov. Accessed 2 Dec 2014
23. New York City Department of City Planning – Population Division, Percent of Persons Below the Poverty Level in 1999 by Census Tract, New York City (2000). http://www.nyc.gov/html/dcp/pdf/census/mpsf3pov2.pdf. Accessed Nov 2014
24. Department of Health and Mental Hygiene (DOHMH). Epiquery: NYC Interactive Health Data System – New York City Health and Nutrition Survey (2012). http://nyc.gov/health/epiquery. Accessed Nov 2014
25. New York City Department of Health Lead Poisoning Prevention Program. Surveillance of Childhood Blood Lead Levels in New York City (2002). https://www.health.ny.gov/environmental/lead/exposure/childhood/surveillance_report/2006-2007/. Accessed Jan 2015
26. New York City Department of Parks and Recreation. Land Cover Raster Data (2010). https://data.cityofnewyork.us/Environment/Landcover-Raster-Data-2010-/9auy-76zt. Accessed Nov 2014
27. New York State Department of Environmental Conservation and New York State Department of Health. New York State brownfield cleanup program, department of soil cleanup objectives (2006). http://www.dec.ny.gov/docs/remediation_hudson_pdf/techsuppdoc.pdf. Accessed Aug 2016
28. New York State Department of Transportation, Traffic Monitoring Unit in the Highway Data Services Bureau. Annual Average Daily Traffic (AADT), 1999–2013 (2013). https://www.dot.ny.gov/tdv. Accessed May 2016
29. Saldivar-Tanaka, L., Krasny, M.: Culturing community development, neighborhood open space, and civic agriculture: the case of latino community gardens in New York City. Agric. Hum. Values **21**, 399–412 (2004)
30. United States Department of Agriculture – Natural Resources Conservation Service (USDA-NCRS), Soil Survey Division. Official Soil Series Descriptions (2007). https://soilseries.sc.egov.usda.gov/osdname.asp. Accessed Mar 2016
31. Schoeneberger, P.J., Wysocki, D.A., Benham, E.C., Soil Survey Staff: Field Book for Describing and Sampling Soils: Version 3.0. Natural Resources Conservation Service, National Soil Survey Center, Lincoln, NE (2012)
32. Sheets, R., Kyger, J., Biagioni, R., Probst, S., Boyer, R., Barke, K.: Relationship between soil lead and airborne lead concentrations at springfield, Missouri, USA. Sci. Total Environ. **271**, 79–85 (2001)
33. Solomon, R.L., Hartford, J.W.: Lead and cadmium in dusts and soils in a small urban community. Environ. Sci. Technol. **10**, 773–777 (1976)
34. United States Census Bureau. Zip Code Tabulation Areas (ZCTAS) (2010). https://www.census.gov/geo/maps-data/data/tiger-data.html. Accessed Jun 2014
35. United States Census Bureau. Population (2010). https://www.census.gov/geo/maps-data/data/tiger-data.html. Accessed Nov 2014
36. United States Environmental Protection Agency (EPA): Lead; Identification of Dangerous Levels of Lead. EPA Federal Register, no. 66, pp. 1206–1240 (2001)
37. Weitzman, M., Aschengrau, A., Bellinger, D., Jones, R., Hamlin, J.S., Beiser, A.: Lead-contaminated soil abatement and urban children's blood lead levels. J. Am. Med. Assoc. **269**, 1647–1654 (1993)

38. Yankel, A., Lindern, I., Walter, S.: The silver valley lead study: the relationship between childhood blood lead levels and environmental exposure. J. Air Pollut. Control Assoc. **27**, 763–767 (1977)
39. Zahran, S., Laidlaw, M.A.S., McElmurry, S.P., Filippelli, G.M., Taylor, M.: Linking source and effect: resuspended soil lead, air lead, and children's blood lead levels in detroit, Michigan. Environ. Sci. Technol. **47**, 2839–2845 (2013)

Organic Matter and Elemental Composition of Humic Acids in Soils of Urban Areas: The Case of Rostov Agglomeration

Sergey N. Gorbov(✉), Olga S. Bezuglova, Svetlana A. Tischenko, and Andrey V. Gorovtsov

Academy of Biology and Biotechnology, Southern Federal University, Rostov-on-Don, Russia
{sngorbov, tischenko}@sfedu.ru, lola314@mail.ru, gorovtsov@gmail.com

Abstract. The aim of investigation was to evaluate the effects of urbanization on the organic matter of urban soils in the South of Russia.

The type of vegetation has an impact on the amount and quality of soil humus. In soils of the forest-park area of the city, an increase of humus content in the upper 10-cm layer to the average value of 7.5 ± 0.63% has been observed, which is significantly higher than the humus content of arable chernozem surrounding the city (3.5–4%).

The humus profile acquires the features of the forest soil. Reduction of humus content is characteristic of buried and sealed soils. The structure and composition of humic acids is largely due to the type of anthropogenic impact. However, regardless of the nature and degree of changes, the transformation of HA molecules are within the defined soil type. The most significant changes are noted for HA from Calcic Chernozem of the forest park. The significant decrease in humus content in urban soils is due to a change of soil conditions. The fact that humic acids of chernozem are less benzenoid and characterized by more advanced peripheral portion of the molecule, with a higher degree of enrichment with nitrogen and sulfur. Sealing the soil under the asphalt leads to the development of HA oxidation and hydrogenation processes. The reduced participation of aromatic moieties in the molecules of HA in the horizon UR of Ekranic Technosol has been noted.

Keywords: Calcic chernozem · Ekranic technosol · Humic acids · Soil organic matter · Urban pedogenesis · Urban soil · Urbic technosol molic

1 Introduction

The study of soil organic matter is a necessary step to solve many issues related to the rational use of soil resources and forecasting the consequences of urban pedogenesis. It is also needed to create the theoretical basis for soil monitoring [13, 15, 17, 29]. The humus substances play a leading role in the formation of a soil body. On one hand, they possess substantial dynamism and capacity to respond to any changes in the environment or to human intervention, and on the other, they are sufficiently "conservative"

V.I. Vasenev et al. (eds.), *Megacities 2050: Environmental Consequences of Urbanization*, Springer Geography, https://doi.org/10.1007/978-3-319-70557-6_10

moieties capable to resist a certain level of disturbing influences, maintaining the ecological status [3, 5, 7, 14, 16, 22, 25, 28, 35].

According to modern concepts, humic acids, the main component of the humus, are a system of high-molecular compounds, and their elemental composition is an average of chemical elements concentration in a group of chemically close, but at the same time, structurally different substances included in the system. However, it was found that the elemental composition of humic acids naturally changes from one type of soil to the other [1].

Orlov [19, 20, 22] has shown that the distribution of carbon in humic acids of the same soil type is approximated by the normal distribution law, with varying rates depending on the type of soil. This allowed him to make a conclusion that humic acid is not a random mixture and its composition depends on the pedogenic conditions. Thus, it was proved that the determination of the elemental composition of humic acid is a necessary stage of research in the study of soil humus status of any soil. There is no doubt, that many other informative methods to study the molecular structure of humic acids exist, but the elemental composition still remains one of the main, and only the significant labor intensity explains the relatively rare use of such a method in the works. However, it is possible to find some information in the literature on the elemental composition of humic acids of various soil and peat types in their natural state [9, 18]. At the same time, such characteristics of technogenically transformed and urban soils are absent. Urbanized areas are characterized by varying degrees of disturbance and anthropogenic transformation of soil [2, 11, 12, 30, 31]. Previously, the significant features of humus composition in urban soils were found [2, 10, 32, 38], therefore the analysis of the elemental composition of HA in these objects also seems to be of particular importance.

Humic acids are substances with irregular chemical composition and structure, characteristic of polydispersity and structural elements heterogeneity [8, 25, 36, 37]. According to Orlov [20], the coefficients of variation of the carbon content in the humic acids obtained from the same type of soil reach 5–8%, which often exceeds the differences between the genetic types. The classic pattern of HA carbon content change in the zonal aspect manifests itself only in average values. His works have shown the average carbon content of the humic acids of main soil types. In particular, in Chernozems the average carbon content is 57.9%, and the variation limits are 52.0−63.8% (in terms of dry ash-free matter).

The aim of the study was to study the composition and structure of humic acids of Chernozem soil under various degrees of anthropogenic impact of the Rostov agglomeration.

2 Materials and Methods

2.1 Description of Field Experiment and Sampling

Rostov agglomeration is a part of the South-Eastern district of Rostov region, occupying about 40% of its territory, and it is strongly monocentric. The total population of the agglomeration is more than 2 million people, the population density – 147 people per square kilometer. Agglomerations are characteristic of an intensive process of adjacent territories development, called suburbanization [34]. As a result, over the past two decades, the city of Rostov-on-Don has de facto merged with Aksay and Bataisk cities.

With consideration for the complex nature of the object under study, 65 soil profiles were established in different regions of the Rostov agglomeration, in which the humus status was studied. The studies were conducted in two periods: 1996–2001 and 2010–2015. When studying the natural soils, the profiles were laid in park and recreational areas of the city and in arable areas adjacent to the edge of the city. As a rule, they are Calcic Chernozem and Calcic Chernozem Aric soil types. The rest of the soil profiles are confined to the industrial and residential areas and represent anthropogenically transformed soils (Ekranic Technosol, Urbic Technosol Molic and Calcic Chernozem Novic Technic). The cultivated black soil (Calcic Chernozem Aric) of the city outskirts and of the Botanical Garden (Southern Federal University), which occupies an area of 160 ha in the geographic center of agglomeration (Fig. 1) were chosen as a comparison object for the artificial forest park soils and for soils with varying degrees of anthropogenic transformation.

In the elemental composition study of the humic acids of urban soils in Rostov agglomeration, we can only refer to previously analyzed humic preparations, but cannot use them as a reference sample. In this regard, aiming for the purity of our experimental design, the preparations of humic acids were made from the horizons of Calcic Chernozem Aric adjacent to the city as well as from urban soils in Rostov-on-Don, identified as the most typical.

Thus, the selected soil types were as follows: (i) Calcic Chernozem, located in the "protective" forest park zone of Rostov-on-Don. This soil is experiencing the impact of woody vegetation uncharacteristic of natural steppe conditions; (ii) Ekranic Urbic

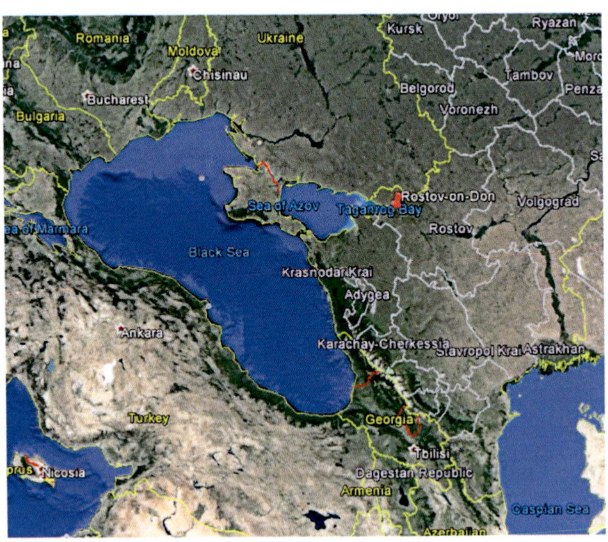

Fig. 1. Locations of soil profiles in Rostov agglomeration.

Technosol Molic, located in the historical center of Rostov-on-Don. This profile combined the asphalt sealed thick urbik horizon and subhorizons with a classic almost full-profiled chernozem, buried at a depth of 115 cm; and (iii) Calcic Chernozem, located in agricultural land within the borders of the agglomeration between the cities of Rostov-on-Don and Aksay. Naturally, arable land was also exposed to human impact, but compared to the urban soils these impacts were minimized and the soil did not experience overt influence of urban pedogenic processes.

2.2 Chemical Analysis

In the study of humus state the determination of organic carbon was conducted twice in the samples from each horizon by two essentially different analytical methods:

1. The wet combustion method was used for the determination of carbon content in accordance with method standardized in the Russian Federation (State Standard 26213-91), 2. Method of high temperature catalytic burning in total organic carbon analyzer TOC-L CPN Shimadzu, where the organic carbon (TOC) is determined as the difference between the overall total (TC) and inorganic (IC) carbon. The principle of total carbon (TC) determination is based on the direct burning of the sample in the TC combustion tube filled with an oxidation catalyst, and heated to 680 °C. Our studies have shown that the results obtained by the oxidation method and by the direct combustion on a Shimadzu TOC-L CPN analyzer are well matched: the differences between the methods lie within the determination error [2].

2.3 Humic Acids Extraction

Extraction of humic acid preparations (HA) was carried out by treating the soil samples with 0.1 N sodium hydroxide after decalcification by 1 N HCl. The isolated raw products were purified by dialysis in cellophane bags, followed by additional purification by electrodialysis. The suspension was then dried at 60 C [24].

Elemental analysis of the obtained HA preparations (i.e., content C, O, H, N) was made in the laboratory of analytical chemistry department, Moscow state university, on CHN-analyzer «Carlo Erba» model 1106, the sulfur content was determined in the Laboratory of Applied Soil Science of the University of Essen (Essen, Germany) on a similar instrument (Carlo Erba). The oxygen in HA preparations were calculated by difference. Elemental analysis data were recalculated to the ash-free sample.

The contents of the carboxylic and hydroxylic functional groups were determined by direct and back titration [24]. The use of a complex of HA isolation and purification methods yielded relatively low-ash products, which allowed to obtain high-quality IR spectra. The IR spectroscopy was carried out in the HA preparations isolated from Ad and A1 horizons of forest park Chernozem, Ap and A of arable Chernozem and UR1 and A_{bur}. of Ekranic Urbic Technosol Molic. The IR spectra of the HA were obtained using KBr technique on a two-beam-automatic infrared spectrophotometer IKS-29 in the range of 4200–400 cm^{-1} [21].

3 Results and Discussion

3.1 Humus State of Urban Soils

The most typical 65 soil profiles were chosen for better representation of the changes in humus state of urban soils. They were combined into two groups depending on the level of transformation of their morphological characteristics:

(i) Natural soils - Calcic Chernozem, Calcic Chernozem Aric
(ii) Anthropogenically transformed soils - Ekranic Technosol, Urbic Technosol Molic and Calcic Chernozem Novic Technic.

The most informative data on total organic carbon content in natural soils and anthropogenically transformed soils of the city are presented in Tables 1 and 2.

According to primary research of the area which is now occupied by the Rostov agglomeration, average (from 27 sections) humus content in the 0–10 cm layer characteristic for Calcic Chernozem of the northern Azov region was 5.7% with a range

Table 1. Contents of total organic carbon in natural soils of Rostov agglomeration

Horizon	Horizon thickness, cm	C org., by wet combustion, %	Humus, by wet combustion, %	TC, %	IC, %	TOC, %	Humus, by TOC, %
1	2	3	4	5	6	7	8
Calcic Chernozem Aric (arable land, Aksay city, settl. Yantarny; profile 1205)							
Ap	0–25	2.62	4.52	2.42	0.01	2.41	4.16
A1	25–40	2.22	3.83	2.38	0.14	2.24	3.87
B1	40–60	1.48	2.55	2.34	0.76	1.58	2.72
B2	60–85	1.14	1.97	2.26	1.18	1.08	1.86
BC	85–110	0.68	1.17	2.30	1.72	0.57	0.99
Cca	110–130	0.57	0.98	2.27	1.81	0.46	0.80
Calcic Chernozem Aric (fallow, Rostov-on-Don city, Botanical Garden (SFedU); profile 1403)							
Ad	0–15	2.22	3.83	2.55	0.02	2.53	4.36
A	15–50	1.83	3.15	2.21	0.01	2.20	3.79
B1	50–65	1.54	2.65	2.13	0.33	1.80	3.09
B2	65–90	1.25	2.16	2.26	1.06	1.19	2.06
BC	90–110	1.03	1.78	2.39	1.88	0.51	0.88
Cca	110–150	0.40	0.69	2.08	1.93	0.15	0.26
Calcic Chernozem (forest park, Rostov-on-Don city; profile 1203)							
Ad	0–10	4.22	7.28	4.16	0.01	4.14	7.14
A1	10–50	2.68	4.62	2.58	0.00	2.58	4.45
B1	50–65	2.34	4.03	1.93	0.01	1.92	3.31
B2	65–90	1.65	2.84	1.29	0.02	1.27	2.18
BC	90–110	0.97	1.67	2.06	1.58	0.48	0.82

Table 2. Contents of total organic carbon in anthropogenically transformed soils of Rostov agglomeration

Horizon	Horizon thickness, cm	C org., by wet combustion, %	Humus, by wet combustion, %	TC, %	IC, %	TOC, %	Humus, by TOC, %
1	2	3	4	5	6	7	8
Urbic Technosol Molic (Rostov-on-Don city, residential part of the city; profile 1303)							
UR1	0–45	1.33	2.29	1.82	0.90	0.92	1.58
UR2	45–70	2.26	3.89	1.78	0.33	1.46	2.51
UR3	70–103	2.45	4.22	1.64	0.39	1.25	2.15
BC	103–130	0.66	1.13	2.05	1.50	0.54	0.94
Cca	130–170	0.50	0.86	1.65	1.32	0.33	0.57
Ekranic Technosol Urbic Molic (Rostov-on-Don city, residential part of the city; profile 1201)							
UR1	18–35	1.26	2.17	–	–	–	–
UR2	35–57	1.03	1.78	1.50	0.83	0.66	1.14
Abur	58–87	1.14	1.97	1.07	0.01	1.06	1.82
B1	87–107	0.97	1.67	1.01	0.01	1.00	1.73
B2	107–133	0.8	1.38	1.73	1.06	0.68	1.17
BC	133–160	0.8	1.38	1.97	1.61	0.36	0.62
Cca	160–206	0.28	0.48	1.61	1.39	0.22	0.38
C	206–280	0.40	0.69	1.37	1.20	0.17	0.29
Ekranic Calcic Chernozem Novic Technic (Rostov-on-Don city, residential part of the city; profile 1401)							
UR1	20–40	–	–	1.95	0.72	1.23	2.11
UR2	40–45	1.2	2.07	1.53	0.13	1.40	2.41
A	45–72	1.37	2.36	1.42	0.01	1.41	2.43
B1	72–92	1.08	1.86	1.02	0.02	1.00	1.73
B2	92–114	0.74	1.28	1.94	1.38	0.56	0.97
BC	114–132	0.51	0.88	2.16	1.92	0.24	0.42
C	132–182	0.28	0.48	1.98	1.83	0.15	0.26
D	182–240	0.06	0.10	0.19	0.05	0.14	0.24

from 4.4 to 7, 6%. Reduction of humus content with the depth was very gradual - from 5.7 to 1.7% at a depth of 100 cm [39].

Studies [10] have shown that an increase of humus content in the forest area is observed in the surface 10 cm layer. During the sixty-year period the humus content of 7.5 ± 0.63% (n = 12, p = 0. 95) have become the average value for these soils varying from 6.5 to 10.0%.

On reserve lands (fallow lands and waste lands) and on agricultural lands surrounding the city the humus content of calcic Chernozems was in average 4.0 ± 0.25% (n = 7, p = 0. 95). At the same time on the fallow plots compared to cropland some increase in humus content of the upper soil horizons has been stated.

Reduction of humus content is a characteristic feature of these soils that is well illustrated by the example of sealed soils [10]. Soils located under the asphalt are usually characterized by lower humus content, for instance in the sealed horizon A, which is covered with an impermeable coating contained only 3.5% of humus.

3.2 The Elemental Composition of Humic Acids

The elemental composition of humic acids, presented in a percent by weight (Table 3) has shown that the carbon content is within these oscillations M $\pm$ σ, determined by Orlov [23]. The findings also confirmed the pattern established for the chernozems of the Rostov region previously: lower carbon content in humic acids as compared to the chernozems in other areas, which is due to the specifics of the provincial humification [4].

The nitrogen content of humic acids changes interestingly through the profile of the studied soils. The enrichment of humic acids with nitrogen is explicitly traced in the upper horizons of the soil under woody vegetation, whereas in arable and buried soils, this phenomenon is not expressed. There was also a slightly higher content of this element in the urbic horizon of Ekranic Technosol.

Humic acids of the sod horizon of Chernozems from park and recreational areas were enriched not only with nitrogen, but also with sulfur and hydrogen.

Table 4 shows the elemental composition of humic acids in terms of atomic percent. The distribution of humic acids carbon in the soil profile has a wavy character. In all three studied soils we noted the high content of this element in the B1 horizon.

Table 3. Elemental composition of humic acids in urbostratozems and chernozems of Rostov-on-Don, recalculated to an ash-free sample

Horizon	Sampling depth, cm	Humic acids; mass percent, %					
		C	H	O	N	S*	ash
Calcic Chernozem (park-recreational zone of the city)							
Asoc	0–10	52.05	4.19	38.96	4.80	0.60	0.56
A	10–25	54.89	3.69	37.21	4.21	0.37	6.02
B1	25–40	54.94	3.50	37.50	4.06	0.37	5.72
B2	40–60	52.65	3.50	39.95	3.90	0.30	3.52
Calcic Chernozem Aric (arable land at the outskirts of the city)							
Ap	0–25	54.48	3.45	38.59	3.48	0.25	3.80
As/p	25–55	54.59	3.34	38.62	3.45	0.22	3.24
B1	55–80	56.69	3.58	36.33	3.40	0.21	2.91
B2	80–105	49.11	3.76	43.53	3.60	0.18	11.40
Ekranic Technosol Urbic Molic (city centre)							
UR	95–115	53.57	4.0	38.4	4.2	0.29	3.47
$A_{bur.}$	115–140	50.74	3.4	42.2	3.7	0.21	2.97
B1	140–160	54.63	3.5	38.3	3.6	0.17	5.17
B2	160–175	52.40	3.2	40.9	3.6	0.20	2.59

Table 4. The elemental composition of humic acids and the degree of their oxidation in the Chernozems and Technosols Rostov-on-Don, recalculated to an ash-free sample

Horizon	Thickness, cm	Elemental composition, at. %				Atomic ratios				Ω
						H:C	O:C	(H:C) Rev.	C:N	
		C	H	O	N					
Calcic Chernozem (park-recreational zone of the city)										
Ad	0–10	38.45	36.86	21.67	3.02	0.96	0.56	1.86	12.71	+0.17
A	10–25	42.12	33.64	21.47	2.76	0.80	0.51	1.48	15.23	+0.22
B1	25–40	42.83	32.52	21.93	2.72	0.76	0.51	1.44	15.76	+0.26
B2	40–60	41.20	32.64	23.52	2.63	0.79	0.57	1.55	15.64	+0.35
Calcic Chernozem Aric (arable land at the outskirts of the city)										
A ar	0–25	42.75	32.20	22.69	2.35	0.75	0.53	1.46	18.16	+0.31
A u/ar	25–55	43.25	31.46	22.91	2.38	0.73	0.53	1.44	18.20	+0.33
B1	55–80	43.83	32.87	21.08	2.23	0.75	0.48	1.39	19.67	+0.21
B2	80–105	37.91	34.48	25.21	2.41	0.91	0.67	1.81	15.73	+0.42
Ekranic Technosol Urbic Molic (city centre)										
UR	95–115	40.07	35.76	21.47	2.70	0.89	0.54	1.61	14.87	+0.18
[A]	115–140	40.27	32.16	25.10	2.48	0.80	0.62	1.63	16.23	+0.45
B1	140–160	42.68	32.46	22.42	2.44	0.76	0.53	1.47	17.50	+0.29
B2	160–175	42.04	30.95	24.69	2.32	0.74	0.59	1.53	18.17	+0.44

In profile distribution of HA nitrogen there is a clear tendency to reduction of the nitrogen content with depth, both in chernozems of recreational part of the city, and in sealed soils.

The oxygen content of humic acids varies. The calculation of the degree of humic acid oxidation by the method of DS Orlov [19] has shown that in all horizons of the studied soils the humic acids are present in reduced form, but the changes in the degree of oxidation through the profile are not the same. Humification process is characterized by an increase of the oxidation degree in the formed products [23], and it is clearly seen in the Chernozems of the recreational area. A gradual increase of this index is noted down the profile with a minimum content in the horizon Ad. At the same time, in sealed and arable analogues the pattern is different: there is a "jump" of the degree of oxidation from the horizon B1 to the horizon B2, where it is much higher than in the other layers.

3.3 The Optical Density of HA

The optical density of the humic acids in studied soils is characterized by the data presented in Table 5. The lowest values of HA optical densities were recorded in the Chernozem of forest park in the surface humus-accumulative horizons Ad and A. At the same time the results of functional groups determination have shown that these horizons are characterized by the highest presence of hydroxylic and carboxylic groups. These low values of the optical density are also observed in the UR horizon of Ekranic Technosol, although high values of neither COOH nor OH groups have been recorded in this horizon during the study.

Table 5. The optical density, color coefficients and the functional groups in Chernozems and Technosols of Rostov-on-Don

Horizon	Sampling depth, cm	0.001% E 1 cm	$E_{465/650}$	$E_{400/500}$	$E_{500/600}$	$E_{600/700}$	COOH	OH
							g-eq/100 g	
Calcic Chernozem (park-recreational zone of the city)								
Asod	0–10	0.07	3.16	2.04	1.85	1.88	0.620	0.360
A	10–25	0.08	2.90	1.89	1.74	1.83	0.632	0.376
B1	25–40	0.11	2.87	1.89	1.74	1.81	0.420	0.179
B2	40–60	0.10	2.84	1.83	1.74	1.83	0.444	0.206
Calcic Chernozem Aric (arable land at the outskirts of the city)								
Ap	0–25	0.10	2.90	1.86	1.74	1.88	0.408	0.235
$A_{s/p}$	25–45	0.12	2.91	1.84	1.74	1.90	0.434	0.235
B1	55–80	0.10	3.04	1.87	1.78	1.97	0.439	0.251
B2	80–105	0.07	3.10	1.89	1.82	1.98	0.300	0.460
Ekranic Technosol Urbic Molic (city centre)								
UR	95–115	0.09	3.04	1.89	1.79	1.92	0.365	0.176
$A_{bur.}$	115–140	0.11	2.63	1.82	1.66	1.73	0.384	0.137
B1	140–160	0.11	2.97	1.88	1.77	1.89	0.424	0.258
B2	160–175	0.10	3.07	1.91	1.79	1.93	0.415	0.193

In the lower, natural horizons the pattern of changes in optical density values of the HA is similar to what is seen in ordinary calcic Chernozem (arable land on the outskirts of the city).

Down through the soil profile the optical density of HA decreases and the color factor increases respectively. The highest E4: E6 value is typical for the sod horizon.

3.4 The IR Spectra of Humic Acids

IR spectra of the studied HA preparations represent a practically identical set of medium intensity bands (Fig. 2), slightly varying in their ratio in the individual soil profiles, with peaks at 3390 (3375), 2925 and 2865, 1710, 1620−1605, 1395, 1245 (1225), 1045, 520 and 465 cm^{-1}. In addition, in the individual spectra ledges, kinks or bands of a very weak intensity are detected at 3195 (3210), 3075, 2550, 1695, 1665, 1650, 1090, 915, 800, 765 cm^{-1}.

Humic acids IR spectra of the two upper genetic horizons of arable Chernozem are almost identical in a set of absorption bands, as well as in their intensity. In the short-wavelength range there is a broad band at 3390 cm^{-1} (stretching vibrations of OH groups of various kinds, linked by hydrogen bonds), in its long-wavelength wing there are bands of a weak intensity at 2926 and 2865 cm^{-1} (stretching vibrations of CH groups in CH2).

In addition, in the horizon Ap a very weak peak at 3225 cm^{-1} may be noted, which is perhaps characteristic of the valent NH group involved in hydrogen associated interactions, and 3075 cm^{-1}, which determines the presence of the aromatic CH groups having no more than 2–3 substituents in the structure. Weak absorption at 2550 cm^{-1} is

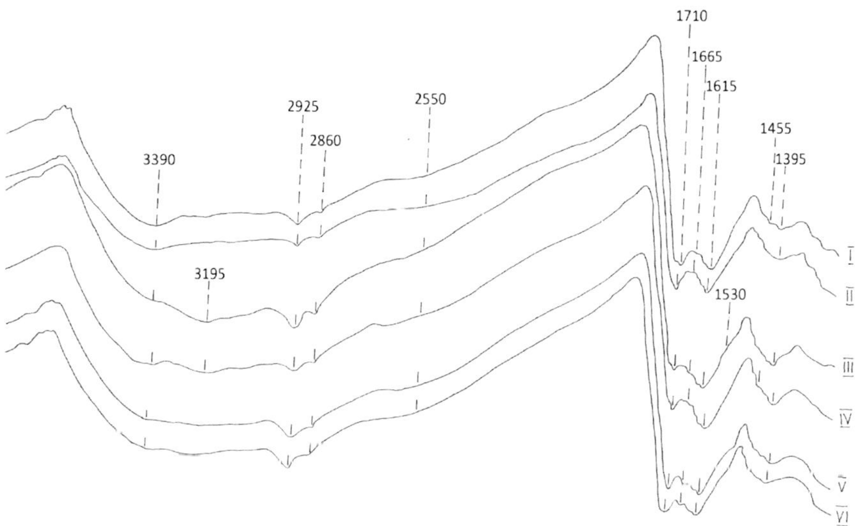

Fig. 2. IR spectra of humic acids of natural and anthropogenically transformed soils of the city (the range of 4200-1200 cm-1): I, II – Ap and Ap1 horizons, (Calcic Chernozem Aric, arable land on the outskirts of the city); III, IV – Ad and A horizons (Calcic Chernozem, park and recreational areas of the city); V, VI – UR and Abur horizons (Ekranic Technosol Urbic Molic, city center).

due to carboxylic acids, more exactly with the OH groups of carboxylic acid dimers. The absorption band of carboxyl groups (C = O of COOH) is comparatively intense and well defined at 1710 cm^{-1} and in the long-wavelength region at 1240 cm^{-1} (C-O in the COOH). Another band at 1610–1620 cm^{-1}, which is associated with the appearance of the C = C benzenoid structures is comparable to the intensity of the absorption band of the carboxyl groups. In the same interval in the humic acids of the Ap horizon there is a weak peak at 1660 cm^{-1} attributable to C = O quinone linked hydrogen bond, and (or) characterizing the expression of C = O groups in the COOH of the aromatic ring.

In the long wavelength region of the spectrum there are absorption bands at 1045–1035 cm^{-1}, 525 and 465 cm^{-1} (Ap horizon), which are due, basically, to the presence of Si-O clay minerals, and in the horizon A1p, there are weak intensity bands at 1095 cm^{-1}, 800 cm^{-1} besides the above-mentioned bands. They are also due to the presence of silicates in the HA samples.

The IR spectra of humic acids isolated from forest park soils were characterized by following differences. The absorption band of carboxyl groups (1710 cm^{-1}) is less intense than in the HA of the arable Chernozem, and it is typical both for HA of the upper and lower horizons of this profile. Furthermore, an increased intensity of various OH groups absorption bands at 3375–3390 cm^{-1}, and possibly also NH-groups, should be noted because a separate absorption band appears in this range at 3195–3210 cm^{-1}, characteristic of the Calcic Chernozem of the forest park.

The appearance of weak ledges at 1650 (Ad) and 1530 cm^{-1} was observed due to the presence of nitrogen-containing groups (amide I and amide II). The intensity of the absorption bands of CH paraffin chains is slightly higher than in the previous profile, but their proportion in the structure of HA is small; the absorption band of benzenoid structures C = C is also clearly expressed at 1605–1620 cm^{-1}.

IR spectra of HA obtained from Ekranic Technosol Urbic Molic U horizon and from the buried horizon (Abur) have no significant distinguishing features in appearance and set of absorption bands when compared to the HA extracted from previous horizons, and are almost identical to each other. They are characterized by a clear absorption band of carboxyl groups (1710 cm^{-1} .1235 and 1245 cm^{-1}, 2550 cm^{-1}) and a weak shoulder at 1660–1665 cm^{-1} (C = O of quinones). C = C bands of the benzenoid structures are also expressed in the same degree regardless of the depth of sampling. The absorption band of CH groups of paraffin chains is slightly increased in the UR horizon. In the HA preparation from the Abur horizon the bands of mineral components (silicates), characterized by at wavelengths 1035 cm^{-1}, 915 cm^{-1} (shoulder), 525 and 465 cm^{-1} appear more intensive. No noticeable manifestation of nitrogen-containing groups in the form of independent highs can be observed, and the absorption band of OH groups (of different kinds) at 3375–3390 cm^{-1} is due to the strong hydrogen bonds, whereby the long-wave wing turns into a plateau.

4 Discussion

4.1 Humus State of Urban Soils

The most of Calcic Chernozem profiles with a high content of organic matter in the surface horizon is confined to the so-called forest park "protective" belt has framing the city of Rostov-on-Don from its eastern and north-eastern sides. This is due to the fact that under the tree vegetation the way in which the plant residues enter the soil is different from the same process in steppe; as a consequence, there is a change in the distribution of humus along the profile. The presence of litter that is produced in forest from fallen leaves and a thick grass cover, protected from the summer fading by the crowns of the trees, is a source of energetic material, enriching the soil with humus and other mineral and organic compounds. As a consequence, humus profile acquires the features of the forest soil: a quite dramatic humus decrease with depth compared to the upper horizons can be noted. Such pattern of humus distribution is characteristic of the nearest northern "neighbor" of Chernozems – dark gray forest soils (Phaeozems). Ponomareva and Plotnikova [27] called a similar distribution of humus in the soil profile "forest type of accumulation of organic matter".

Such an increase in the concentration of organic matter in soils of Moscow was mentioned in the works of Prokofieva and others [31], confirming the fact that similar processes are not unique to urban soil of southern Russia.

Thus, the forest-park "protective" plantings at the outskirts of the city ensuring the creation of wind barrier, and thus the favorable climate of the city are also a dynamic environment-forming factor, leading to the formation of "islands" of a totally different type of matter redistribution in the plant-soil system on a background of steppe nutrient cycle. The tree plantations in steppe are the centers of more intense cycle of matter than the natural steppe communities, and even more intense when compared to arable land.

At the same time, on the fallow plots compared to the cropland some increase in humus content of the upper soil horizons has been noted. When plowing destroys the natural vegetation, the annual flow of all biomass created during the vegetation period stops, and there is a change of redox regime. As a result of this the organic matter changes qualitatively.

Regarding anthropogenically transformed soils, the processes of burial and sealing practically cease the flow of the fresh organic matter to the soil, contemporary processes of humification are damped and, as a consequence, there is no update of humic substances [10]. As a consequence, there is a significant change in the carbon and nitrogen cycles dynamics in the buried soil stratum [17, 26]. In addition, a large proportion of contaminated sediments bypasses soil body, and all this leads to the fact that the possibilities for the soil and humus to conduct their protective features are sharply reduced. Such sealed soils can be regarded as anthropogenic models, reflecting the effect of time on the functioning and the status of soil humic substances system.

Buried soil, unlike sealed under the asphalt (Ekranic Technosol), and does not completely lose contact with the ground surface. Although they lack the influence of many factors of soil formation, the still don't lose their protective function due to partial permeability of the upper strata.

In our study of urban sealed and buried soils significant changes in the content and composition of humus have been found, which is associated with changes in soil conditions and is manifested primarily in the restructuring of the soil profile. Reduction of humus content in Ekranic Technosol and Urbic Technosol Molic is quite clear: the link between soil and plants is disrupted. As a result, the cycling of matter inherent to the steppe zone ceases or changes dramatically.

4.2 The Elemental Composition of Humic Acids

Enrichment of humic acids from park and recreational area Chernozem Ad horizon with nitrogen and hydrogen indicate the relative immaturity of these acids, and significant involvement of peripheral fragments in the structure of their molecules. This fact is explained by the constant updating of humus due to the fresh organic matter entering the soil from the litter of woody plants.

In the arable soil and in buried horizons their amount of fresh organic material is lower. As a result, the mass percent of nitrogen is in the adjacent ranges within both soil types. The exception is urbic horizon, where the mass percent of nitrogen is 4.2%, which may be at part indicative of its anthropogenic origin.

The expression of the elemental composition in mass percentages does not give a complete picture neither of the role of any individual element in the structure of the substance, nor about the changes that occur with humic substances in the soil [22]. Conversion to atomic percentages reveals division between gray and brown humic acids, confirmed by calculations of the oxidation degree, and the analysis of atomic ratios diagram by Van Kleveren method gives the understanding of the direction of the humification process [20].

Increased carbon content in the B1 horizon of the studied soils indicates a specificity of molecular structures at these depths, or rather the dominance of an aromatic central part in their structure. This can be judged more exactly by the ratio (H: C) Rev. (Table 4).

It's common known that the decrease of the H:C ratio indicates an increase in the proportion of aromatic moieties in the molecules of humic acids. But this index is not completely true due to the effect of unaccounted oxygen atoms in the molecule of humic acid, so it is more correct to use the ratio (H:C). Rev., calculated on the basis of the oxygen function in the manner proposed by DS Orlov [22]. The (H:C). Rev. value is calculated as follows:

(H:C). Rev. = (H:C) + 2* (O:C) * 0,67 where (H:C) is the value found through elemental analysis, 2 is the quotient showing that one oxygen atom can substitute 2 hydrogen atoms, i.e. in a carbonyl group, and 0,67 is a quotient taking into account the presence of different oxygen containing groups, including those that substitute only one hydrogen atom, i.e. hydroxylic groups.

Result of elemental analysis and the ratio (H:C) Rev. have shown that the transformation of humic acids depend on the type of anthropogenic impact. The results show a decline of the ratio (H:C) Rev. in the B1 horizon of all studied soils. In other words, whatever changes is the soil exposed to, the accumulation of more benzenoid HA molecules is always marked at the level of carbonate barrier. The most developed aliphatic part of the HA molecules is found in sod horizon, which is clearly due to constant supply of fresh plant residues and thus "immaturity" of the carbon skeleton of the newly formed HA, containing a higher proportion of hydrogen, nitrogen and sulfur. For the same reasons humic acids obtained from Chernozem of the recreational area of the city stand out for the degree of enrichment with nitrogen.

This is consistent with the chemical analyzes of soil, according to which the ratio of gross amounts of C:N is characterized by a somewhat smaller values.

The tendency to the reduction of the nitrogen content with depth, both in chernozem of the recreational part of the city, and in sealed soils is due to the fact that in the upper horizons the plant residues supply (or supplied, as in the case of sealed profiles) much more nitrogen than the lower layers, and in the process of humification this nitrogen is "captured" by humic acids.

The increased oxidation degree, and the positive ω index can serve as an evidence of the most favorable conditions of humification at these depths and leaching of carbonates get some contributes to this process (3). The mentioned data is confirmed by graphical statistical analysis by Van Krevelen method (Fig. 3).

The analysis has shown that the most significant differences are observed in the surface layer of soil. Comparing the Ap, Ad and Abur horizons has shown that woody vegetation is accompanied by a process of methylation of humic acids, which results in the predominance of the peripheral part of the HA.

Sealing the soil leads to the development of oxidation and hydrogenation: the point 2 on the chart (Abur) shifts along the HA molecule hydration line.

The latter is even more characteristic for the plow horizon of the soil of agricultural lands surrounding the city. Humic acids extracted from subsurface horizons A and B1 of the studied soils, are different from each other to a much lesser extent, and this also applies to the buried profile of Chernozem.

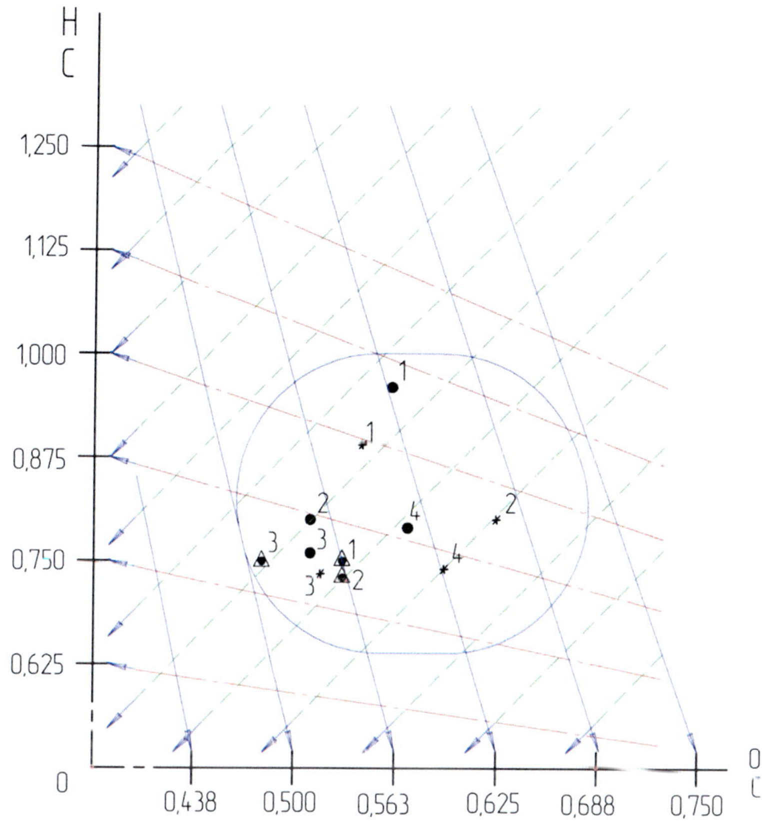

Fig. 3. An analysis of the elemental composition of humic acids by Van Krevelen method:

Δ - Calcic Chernozem Aric (arable land on the outskirts of the city):
1 - Ap horizon; 2 - Ap1 horizon; 3 - B1 horizon;
• - Calcic Chernozem (park and recreational area of the city):
1 - Ad horizon; 2 - A horizon; 3 - B1 horizon; 4 - B2 horizon;
* - Ekranic Technosol, (city center):
1 - UR horizon; 2 - Abur horizon; 3 - B1 horizon; 4 - B2 horizon;

←——— - The direction of demethylation;
←— — — - The direction of dehydration;
←- - - - The direction of decarboxylation.

4.3 The Optical Density of HA

Electronic absorption spectra and extinction coefficients have become widely used to study the nature of humic substances and they are their diagnostic sign [6]. According to modern concepts the color of humic acids and thus the nature of their electronic spectra are caused by a developed system of conjugated double bonds: these are multiple carbon-carbon (ethenyl) and carbon-oxygen (carbonyl) communications that are part of the so-called chromophore compounds.

The land use has left its mark on the properties of structural condition of humic acid. The most significant changes have been noted for the HA molecules of the chernozem of the forested area. Along with the lowest values of optical densities in the upper horizons, lower values of this indicator were found throughout the soil profile. Plant residues which are not typical (in their composition) for the steppe area and the migration of newly formed humic substances down the profile lead to the fact that humic acids have less condensed nucleus and thus more developed peripheral portion of the molecule. The lowest level of condensation in the HA molecules from the horizon Ad is due to the presence of newly formed humic acids, which in turn is connected with a constant supply of fresh organic material. This is also evidenced by the results of the functional groups determination – in this horizon of soil under woody vegetation the highest content of hydroxylic and carboxylic groups was found, indicating greater reactivity of humic acids and their relative immaturity.

The UR horizon of the Ekranic Technosol also has lower optical density and high E4:E6 ratio value, close to that of Ad horizon. But taking into account the low quantity of hydroxylic and carboxylic groups in the structure of HA this is rather the evidence of the anthropogenic origin of the horizon. Down the soil profile the optical density of the HA decreases and the chrominance coefficient, respectively, increases.

Thus, the more thermodynamically stable humic acids correspond to higher optical density, lower values of the chrominance coefficient and fewer functional groups.

In his works Salfeld [33] recommends to determine the three extinction coefficients - E400/E500, E500/E600, E600/E700. The spectrum is represented as a point in a three-dimensional coordinate system that allows getting general information about the correlation of the spectra in the system of humic substances of different origin. After these measurements, we found out that, no matter how strong would be the changes in the HA structure of the investigated soils, they still follow the path of transformation, that is typical for the chernozems. This is clearly shown in the Fig. 4.

4.4 The IR Spectra of Humic Acids

All studied humic acids are characterized by high intensity of the absorption bands of carboxyl groups, the benzenoid structures at 1610–1620 cm^{-1} are also clearly manifested.

The increase in the proportion of OH groups (alcohols, possibly phenols), CH groups of the paraffin chains and nitrogen-containing groups (HA of the recreational area of the city chernozems stand out for their nitrogen richness) indicates a higher enrichment of these HA with peripheral aliphatic chains. This is consistent with the data of elemental analysis of humic acids, presented earlier in this article, and is connected, in our opinion, to the characteristics of the fresh organic matter inflow and transformation.

Humic acids extracted from Ekranic Technosol horizons, are very close in their IR spectra to HA of the arable chernozem only to mention a few less intense absorption bands linked to the COOH groups.

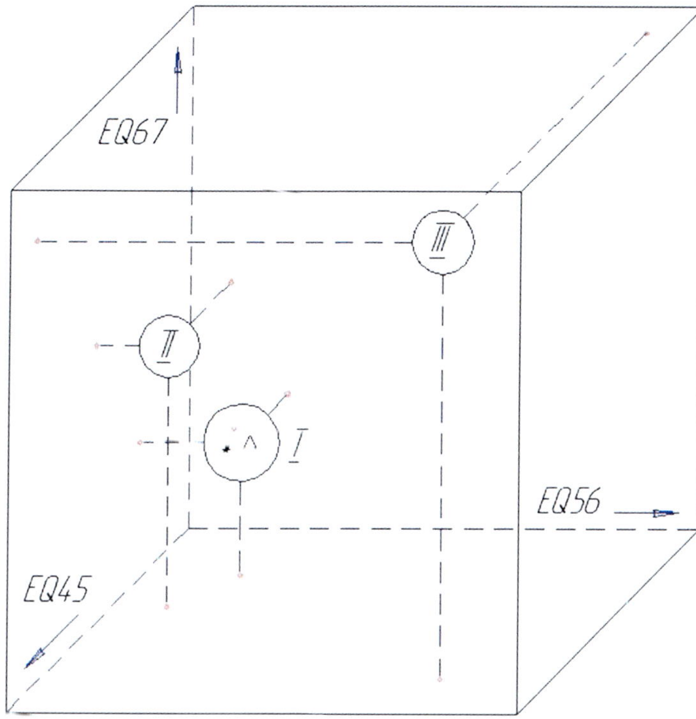

Fig. 4. Schematic model of the imaging the optical density of humic acids using three extinction coefficients.: I – Humic acids of chernozems (•, Δ) and anthropogenically transformed soil of the city (*); II – Humic acids of podzolic soils (from Zalfeld); III – Humic acids of other soil types (from Salfeld, [33]);

5 Conclusions

In the Ekranic Technosol and Urbic Technosol Molic, a significant transformation in the humus content has been found associated with changes in pedogenic conditions and is manifested primarily in the restructuring of the soil profile. Reduction of humus content is a characteristic feature of these soils, since the link between soil and plants is broken. As a result, the matter cycling typical for the steppe zone ceases or changes dramatically.

In soils of the forest-park area of the city, an increase of humus content in the upper 10-cm layer to the average value of 7.5 ± 0.63% is observed, which significantly exceeds the values of humus content of arable chernozem in the agricultural lands surrounding the city (3.5–4%). The humus profile acquires the features of the forest soil, and as a consequence, a quite dramatic decrease of humus content with depth has been observed.

Specific traits of humus formation in chernozems associated with different kinds of changes in their natural cycle of matter, was reflected in the elemental composition of humic acids. However, in general, humic acids are very stable system, and even under a very strong anthropogenic impact the changes are not so significant. The changes in the structure of the HA from the studied soils goes on the way that is typical for chernozems.

The most significant changes are noted for HA molecules from the chernozem of the forest park. The plant residues that is not typical for the steppe zone in the quantity and composition lead to the fact that humic acids of these soils are characterized by a smaller aromatic portion of the molecule, and consequently, a more developed peripheral part. The transformation of plant residues is accompanied by a process of methylation of humic acids. Sealing of the soil leads to the development of oxidation and hydrogenation processes. The optical density of the HA from the studied soils is within the limits inherent to the chernozem zone. The most significant changes are noted for HA molecules from the soils of forested area. All HA preparations are characterized by high intensity of the absorption bands of carboxyl groups and a clear manifestation of the benzenoid structures bands at 1610–1620 cm^{-1}. HA of the forest park chernozem are characterized by an increased proportion of the OH groups (alcohols, possibly phenols), CH groups of the paraffin chains and nitrogen-containing groups, which may indicate a higher enrichment of HA with peripheral aliphatic chains. Humic acids extracted from Ekranic Technosol soil horizons are very close in their IR spectra to HA of the arable chernozem only to mention a few less intense absorption bands linked to the COOH groups.

Acknowledgments. This research was supported by project of Ministry Education and Science of Russia, no. 6.6222.2017/BP. Analytical work was carried out on the equipment of Centers for Collective Use of Southern Federal University "Biotechnology, Biomedical and Environmental Monitoring" and "High Technology".

References

1. Barančíková, G., Senesib, N., Brunetti, G.: Chemical and spectroscopic characterization of humic acids isolated from different Slovak soil types. Geoderma **78**, 251–266 (1997)
2. Bezuglova, O.S., Gorbov, S.N., Tischenko, S.A., Aleksikova, A.S., Tagiverdiev, S.S., Sherstnev, A.K., Dubinina, M.N.: Accumulation and migration of heavy metals in soils of the Rostov region, South of Russia. J. Soils Sediments **16**, 1203–1213 (2016)
3. Bezuglova, O.S., Gorbov, S.N., Privalenko, V.V.: The humus profile and the microelemental composition of soils in the recreational areas of Rostov-on-don. Eur. Soil Sci. **33**, 1001–1006 (2000)
4. Bezuglova, O.S., Zvyagintzeva, Z.V., Goryainova, N.V.: Humus losses in soils of the Rostov province. Eur Soil. Sci. **28**, 40–53 (1996)
5. Chabbi, A., Rumpel, C., Kögel-Knabner, I.: Stabilised carbon in subsoil horizons is located in spatially distinct parts of the soil profile. Soil Biol. Biochem. **41**, 256–261 (2009)
6. Chen, Y., Senesi, N., Schnitzer, M.: Information provided on humic substances by E4/E6 ratios. Soil Sci. Soc. Am. J. **41**, 352–358 (1977)
7. Chukov, S.N.: Structural-functional parameters of soil organic matter under anthropogenic impact. Publishing house of St. Petersburg state University, Russia (2001) (in Russian)

8. Dergacheva, M.I., Gavrilov, D.A., Ochur, K.O., Nekrasova, O.A., Okoneshnikova, M.V., Vasileva, D.I., Ondar, E.E.: Ratio of elements in humic acids as a source of information on the environment of soil formation. Contemp. Probl. Ecol. **5**, 497–504 (2012)
9. Gondar, D., Lopez, R., Fiol, S., Antelo, J.M., Arce, F.: Characterization and acid–base properties of fulvic and humic acids isolated from two horizons of an ombrotrophic peat bog. Geoderma **126**, 367–374 (2005)
10. Gorbov, S.N., Bezuglova, O.S.: Specific features of organic matter in urban soils of Rostov-on-Don. Eur. Soil Sci. **47**, 792–800 (2014)
11. Gorbov, S.N., Bezuglova, O.S., Varduni, T.V., Gorovtsov, A.V., Tagiverdiev, S.S., Hildebrant, Y.A.: Genotoxicity and contamination of natural and anthropogenically transformed soils of the City of Rostov-on-Don with heavy metals. Eur. Soil Sci. **48**, 1383–1392 (2015)
12. Greinert, A.: The heterogeneity of urban soils in the light of their properties. J. Soils Sediments **15**, 1725–1737 (2015)
13. Gruneberg, E., Schoning, I., Kalko, E.K.V., Weisser, W.W.: Regional organic carbon stock variability: a comparison between depth increments and soil horizons. Geoderma **155**, 426–433 (2010)
14. Kholodov, V.A., Konstantinov, A.I., Belyaeva, E.Y., Perminova, I.V., Kulikova, N.A., Kiryushin, A.V.: Structure of humic acids isolated by sequential alkaline extraction from a typical chernozem. Eur. Soil Sci. **42**, 1095–1100 (2009)
15. Kögel-Knabner, I.: Analytical approaches for characterizing soil organic matter. Org. Geochem. **31**, 609–625 (2000)
16. Kögel-Knabner, I.: The macromolecular organic composition of plant and microbial residues as inputs to soil organic matter. Soil Biol. Biochem. **34**, 139–162 (2002)
17. Lorenz, K., Kandeler, E.: Biochemical charaterization of urban soil profiles from Stuttgart, Germany. Soil Biol. Biochem. **37**, 1373–1385 (2005)
18. Martin, D., Srivastava, P.C., Ghosh, D., Zech, W.: Characteristics of humic substances in cultivated and natural forest soils of Sikkim. Geoderma **84**, 345–362 (1998)
19. Orlov, D.S.: The Elemental composition and the degree of oxidation of humic acids. Biol. Sci. **1**, 5–20 (1970). (in Russian)
20. Orlov, D.S.: Modern chemical and physical methods of investigating the nature and structure of humic substances of soils. Soil Sci. **7**, 55–62 (1972). (in Russian)
21. Orlov, D.S., Osipova, N.N.: Infrared spectra of soils and soil components. Moscow University Press, Moscow (1988). (in Russian)
22. Orlov, D.S.: Humus Acids of Soils and General Humification Theory. Moscow University Press, Moscow (1990). (in Russian)
23. Orlov, D.S.: Soil Chemistry. Oxford and IBH Publishing Co, NewDelhi (1992). pp. 213–300
24. Orlov, D.S., Grishina, L.A.: Practical Manual of Humus Chemistry. Moscow University Press, Moscow (1981). (in Russian)
25. Perminova, I.V., Frimmel, F.H., Kudryavtsev, A.V., Kulikova, N.A., Abbt-Braun, G., Hesse, S., Petrosyan, V.S.: Molecular weight characteristics of aquatic, soil, and peat humic substances as determined by size exclusion chromatography and their statistical evaluation. Environ. Sci. Technol. **37**, 2477–2485 (2003)
26. Piotrowska-Długosz, A., Charzyński, P.: The impact of the soil sealing degree on microbial biomass, enzymatic activity, and physicochemical properties in the Ekranic Technosols of Toruń (Poland). J. Soils Sediments **15**, 47–59 (2015)
27. Ponomareva, V.V., Plotnikova, T.A.: Humus and Soil Formation. Leningrad (1980) (in Russian)
28. Popov, A.I.: Humic substances: properties, structure and formation. Publishing House of St. Petersburg State University, Russia (2004)

29. Pouyat, R., Groffman, P., Yesilonis, I., Hernandez, L.: Soil carbon pools and fluxes in urban ecosystems. Environ. Pollut. **116**, 107–118 (2002)
30. Prokof'eva, T.V., Gerasimova, M.I., Bezuglova, O.S., Bakhmatova, K.A., Gol'eva, A.A., Gorbov, S.N., Zharikova, E.A., Matinyan, N.N., Nakvasina, E.N., Sivtseva, N.E.: Inclusion of soils and soil_like bodies of urban territories into the russian soil classification system. Eur. Soil Sci. **47**, 959–967 (2014)
31. Prokof'eva, T., Gerasimova, M., Lebedeva, I., Martynenko, I.: An attempt of integrating the systematic of urban soils into the new Russian soil classification system. Soil Sci. Ann. **64**, 24–28 (2013)
32. Prokof'eva, T.V., Rozanova, M.S., Poputnikov, V.O.: Some features of soil organic matter in parks and adjacent residential areas of Moscow. Eur. Soil Sci. **46**, 273–283 (2013)
33. Salfeld, J.C.: Optical measurements on human systems sympos «Humus et Planta, V» , Prague (1971)
34. Salvati, L., Ferrara, C., Ranalli, F.: Changes at the fringe: soil quality and environmental vulnerability during intense urban expansion. Eur. Soil Sci. **47**, 1068–1075 (2014)
35. Schmidt, M.W.I., Skjemstad, J.O., Gehrt, E., Kögel-Knabner, I.: Charred organic carbon in German chernozemic soils. Eur. J. Soil Sci. **50**, 351–365 (1999)
36. Tikhova, V.D., Fadeeva, V.P., Dergacheva, M.I., Shakirov, M.M.: Analysis of humic acids from various soils using acid hydrolysis. Russ. J. Appl. Chem. **81**, 1957–1962 (2008)
37. Tikhova, V.D., Shakirov, M.M., Fadeeva, V.P., Dergacheva, M.I., Kallas, E.V., Orlova, L. A.: Elemental content and functional group analysis of humic acids in fossilized soils of different ages. Russ. J. Appl. Chem. **74**, 1380–1384 (2001)
38. Vasenev, V.I., Prokofeva, T.V., Makarov, O.A.: The development of approaches to assess the soil organic carbon pools in megapolises and small settlements. Eur. Soil Sci. **46**, 685–696 (2013)
39. Zakharov, S.A.: Soils of the Rostov region and their agronomic characteristics. Rostov-on-Don (1946) (in Russian)

The Effects of Soil-Ecological Factors on the Pb Migration in the Soil of Urban Forest Ecosystem

Elvira Anvarbekovna Dovletyarova[1(✉)], Lyudmila Vladimirovna Mosina[2], Ramilla Adim kizi Hajiaghayeva[1], Nailya Vafaevna Aliyeva[1], and Vadim Gennadyevich Plyuschikov[1]

[1] Agrarian-Technological Institute, RUDN University, Miklukho-Maklaya Str., 8/2, Moscow 117198, Russia
elvira_2779@mail.ru
[2] Moscow Timiryazev Agricultural Academy, Russian State Agrarian University, Timiryazevskaya Str., 49, Moscow 127550, Russia

Abstract. Leadisone of the most dangerous heavy metals for human. The studies of lead migration in the soil as are significantly important to forecast the transfer factors of lead in ecosystems. The paper represents data of lead concentration in dependence of the soil depth and degrees of anthropogenic impact. The research revealed influence of anthropogenic impact and species of arboreal vegetation on Pb migration in the soil profile. The results of lead concentration positively varyin the soil due to anthropogenic impact from 62.2 mg/kg to 139 mg/kg. The effect of arboreal vegetation species results in the high Pb migration down the soil profile in pine and birch sites and preferable accumulation of Pb in the upper humus horizon in the oak sites.

Keywords: Heavymetals · Lead (Pb) · Soil-ecological parameters · Anthropogenic impact

1 Introduction

Accumulation of heavy metals (HM) in the upper humus layer is reported by many studies [3, 8, 9]. Migration of HM is limited in the lower horizons [1]. The main factors, limiting Pb migration to subsoil are the following: high content of SOM (soil organic matter), the neutral soil pH, the optimal hydrological conditions and redox (RH) conditions [18]. The precipitation of heavy metals and metalloids in the soils of south-eastern district of Moscow on the geochemical barriers has the following regularities: Cd, Cu, and Zn are accumulated on the alkaline barriers; Bi, Sb, As, Cu, Pb, and Zn - on chemisorption barriers; Sb, As, and Pb - on organomineral barriers; and Cd and Cu, on the sorption-sedimentation barriers [10].

In addition to the soil-ecological parameters, the HM migration capacity is determined by the metal. Mercury and zinc are characterized with the highest migration

V.I. Vasenev et al. (eds.), *Megacities 2050: Environmental Consequences of Urbanization*, Springer Geography, https://doi.org/10.1007/978-3-319-70557-6_11

ability with soil depth, while the lead and cadmium are less mobile in the soil [4, 5]. Hence concentration of lead is higher in the upper humus layer, closer to the surface of the contaminated soil. However, in highly contaminated soils elevated concentrations of lead at the 40 cm depth are also found [16].

Complexes of lead with humic acids are about 150 times stronger than similar complexes with cadmium [6]. Most of available studies ob HM migration focus on urban or agricultural soils. The data for forest soils is limited and for the Moscow Region such information is lacking. Meanwhile, migration capacity might be different in forest soils, in consequence of deep penetration of the root system.

Arboreal vegetation, characterized with strong and deep root system are able to transport HM into the deeper soil horizons, transforming the soil as a source of toxic substances for many years [13]. The heterogeneity of the soil profile contributes the change of HM availability for plants, increasing risk of HM entering in the food chain [11, 12]. Different micro and macro- element soil composition is the reason of the different HM interaction effect – the synergistic or antagonistic, affects the state of the plants, reducing or increasing their stability [2, 15].

Lead is the one of the most dangerous HM for human. Proceeding through the food chain in the human organism, the excess of lead suppresses the central nervous system, brain function, kidney, muscle; adversely affect the processes of hematopoiesis [7, 17]. The exceptional danger of lead is entering in the food chain and distribution in the urban functioning and suburban forest ecosystems, deteriorating the sanitary functions. The information of lead behavior in the forest experimental station (FES) of Moscow Agricultural Academy soils is practically absent. The research aimed to study the behavior and distribution of lead in the FES of Moscow Agricultural Academy soils.

2 Materials and Methods

Soil samples (HaplicLuvisolaccording to IUSS, 2014, 60 samples) were collected from the experimental plots (the size of 0.5 ha), in the FES located in the northern district of the Moscow city (Russia) on the watershed plateau, relatively far from the influence of industrial facilities and highways (450 to 500 m apart) [13].

The living conditions of the stands are studied as a risk criterion (J) determined by degradation factors. The high value of J shows high anthropogenic impact. In the research 3 sites with different degrees of J are studied. The Ovalue was 2.8 on the verge of disintegration (the Oak, X–XII); 2.5 – weakened condition of plant (the Oak, VII–VIII) and 1.5 – healthy state (the Pine with the birch, IX–XI) [14].

To measure the effect of unregulated recreation, bulk density and soil density were sampled every season from the 0–3 cm, 3–7 cm, and 7–11 cm depth intervals as well as from a full 0–10 cm depth profile. Compaction of the soil layers determines water-air and temperature regimes, redox conditions, and biochemical processes [19]. Lead content in our research was measured by mass-spectrometry.

3 Results and Discussion

Results of the lead distribution in the studied soil profiles are presented at the Figs. 1, 2 and 3. Observed accumulative-eluvial-illuvial character of Pb distribution is characterized by maximum accumulation of the element in the surface layer of soil, and sufficiently deep migration into the soil.

The Pb content correlated with the degree of anthropogenic impact on the surface layer of soil and varied from 62.2 mg/kg (the Oak, X–XII) to 139 mg/kg (the Pine with the birch, IX–XI). Topsoil contents of Pb exceeded subsoil contents 3–10 times. In the illuvial horizon the maximum of the Pb accumulation was observed in the Pine with the birch, IX–XI site. The results revealed significantly dependence of Pb depth migration with the degree of anthropogenic impact.

Lead migration into the soil was positively correlated with soil compaction. At sites of natural forests and remote from the city highways with 0.6–0.8 g/cm^3 soil density content of total Pb (average data) in the illuvial horizon about 2–3 times lower than in sites with increased anthropogenic impacts.

Maximum lead accumulation in the surface layer of the soil indicates the aerial character of contamination. The depth of Pb migration into the lower soil horizons, as well as the scale of distribution in the soil profile is determined by the composition of arboreal vegetation. E.g., in the pine with the birch site Pb content in the B soil horizon is 42 mg. Under the similar conditions with the same distance from the highways under the oak forest, Pb content in the horizon ranges from 18 to 26 mg. Therefore the level of lead pollution at the B horizon of pine with the birch site in 1.6–2.5 times higher

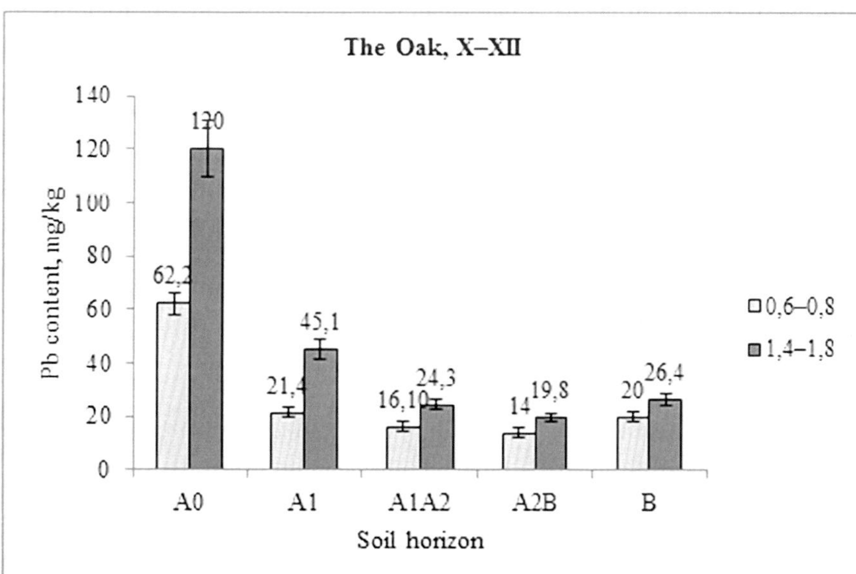

Fig. 1. Distribution of lead in the soil profile with J 2.8.anthropogenic impact degree (verge of disintegration)

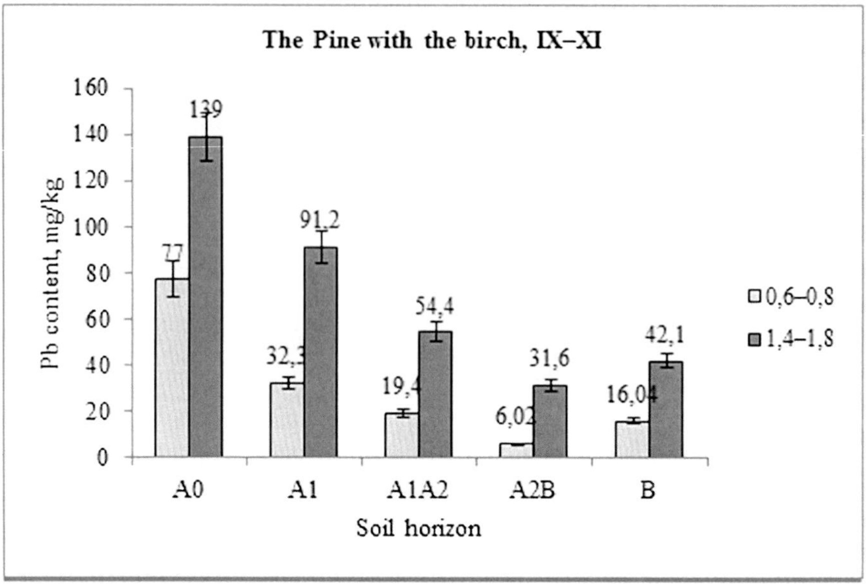

Fig. 2. Distribution of lead in the soil profile with J 2.5.anthropogenic impact degree (weakened condition of plant)

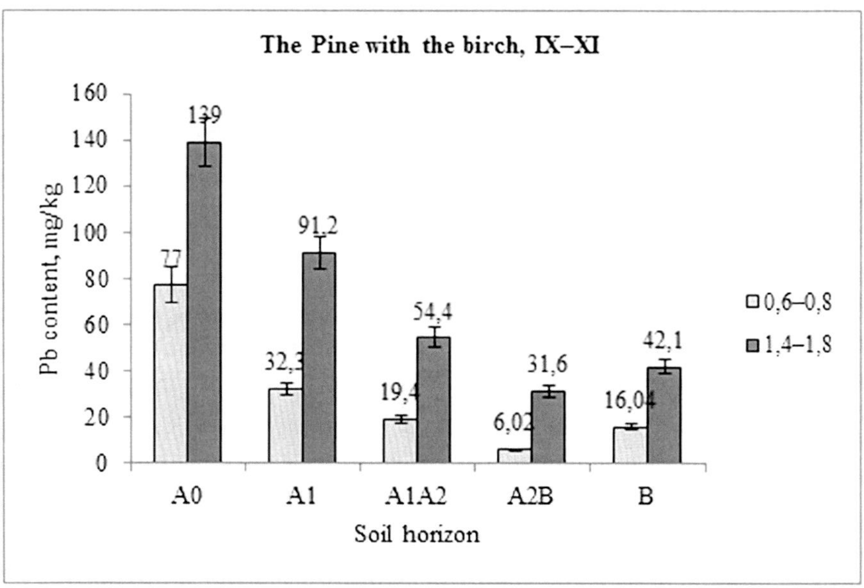

Fig. 3. Distribution of lead in the soil profile with J 1.5.anthropogenic impact degree (healthy state)

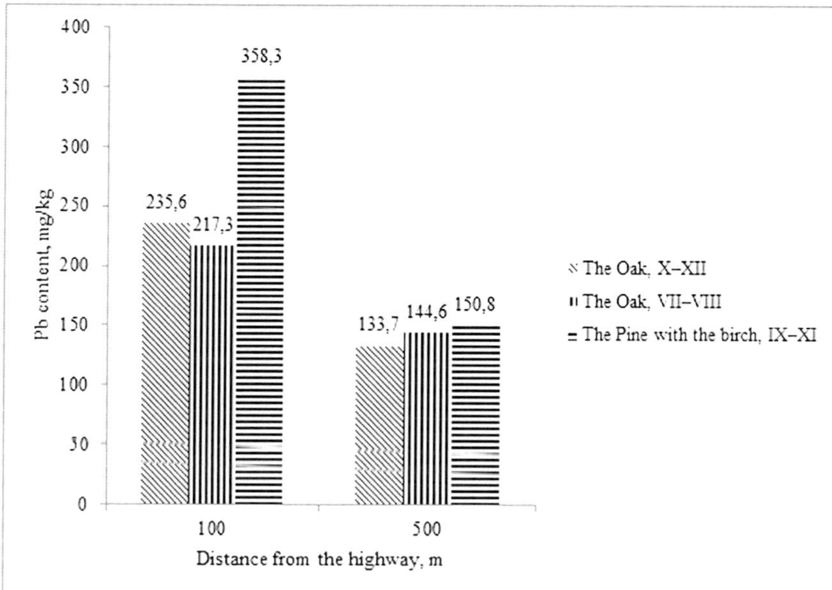

Fig. 4. Accumulation of lead (total content) in the soil profile, depending on the distance from the highway, mg/kg

Table 1. Distribution of the lead in the soil profile under various anthropogenic impacts per 1 cm (1 – Soil density 0.6–0.8 g/cm^3, 2 – Soil density 1.4–1.8 g/cm^3).

Experimental site	Gentichorizon	Depth of sampling, cm		The content of Pb, mg/kg 1 cm layer	
		1	2	1	2
The Oak, X–XII	A_1	0–10	0–4	6.2	30.0
	A_1^1	10–20	4–22	2.1	2.5
	A_1A_2	28–32	30–40	4.0	2.4
	A_2B	45–55	55–65	1.4	2.0
	B	70–80	96–102	2.0	4.4
The Oak, VII–VIII	A_1	0–12	1–6	7.2	22.8
	A_1^1	12–30	6–23	1.5	2.9
	A_1A_2	35–45	25–30	1.5	4.0
	A_2B	55–60	40–50	1.0	1.5
	B	98–104	70–80	2.0	1.8
The Pine with the birch IX–XI	A_1	0–6	0–6	12.8	23.2
	A_1^1	6–24	6–21	1.8	6.0
	A_1A_2	30–40	23–30	1.9	10.8
	A_2B	50–55	40–46	1.2	5.5
	B	76–83	60–70	2.3	4.2

than in the soils of oak forests in increased anthropogenic impact conditions. Perhaps the high content of Pb at the considerable depth of the soil layer under these plantations is due to strong and deep root system, contributes to the more intense movement of Pb in the deeper layers of the soil. In general, the accumulation of lead (total content) in the root layer of soil (to a depth of illuvial horizon B) is significantly higher at sites of the forest, close to the urban thoroughfares (Fig. 4).

In the Table 1 lead distribution down the soil profile under various anthropogenic impacts is showed with the metal concentration expressed per 1 cm layer, associated with different depth of soil genetic horizons.

Per 1 cm root zone Pb content was 1.5–3 times higher in forest areas with high anthropogenic impact (close to urban thoroughfares and in conditions of high recreational load). In this case the maximum values (5.1 mg) of Pb concentration reaches 1 cm root zone soil layer under the pine and birch phytocenosis.

4 Conclusions

The obtained results show significant accumulation of lead in the upper humus layer under increased anthropogenic impact conditions, confirming aerial character of contamination. In the upper humus horizon (A_1 horizon) the accumulative effect is 2–5 times higher in compacted soils compared to less compacted soils. The accumulative effect was 22.8 and 7.2 mg for single-species stands and 23.12 and 12.8 for soils under pine and birch phytocenosis respectively. Pine and birch stands are active conductors of Pb in the soil profile; oak phytocenoses accumulate greater extent of Pb in the upper humus horizon In the upper humus horizon (A_1) observed differences in the accumulative effect are approximately 1.5–2 times in the forest areas with high anthropogenic impact.

The study reveals the possibility of anthropogenic impact influence and species of arboreal vegetation on Pb migration in the soil profile. The results allow forecasting lead mobility in the urban and suburban landscapes and provide an opportunity to develop recommendations for the urban land using of the areas affected by lead contamination.

Acknowledgements. The publication was prepared with the support of the "RUDN University program 5-100".

References

1. Andráš, P., Turisová, I., Krnáč, J., Dirner, V., Voleková-Lalinská, B., Buccheri, G., Jeleň, S.: Hazards of heavy metal contamination at L'ubietová cu-deposit (Slovakia). Procedia Environ. Sci. **14**, 3–21 (2012)
2. Cheng, S.: Effects of heavy metals on plants and resistance mechanisms. Environ. Sci. Pollut. Res. **10**, 256–264 (2003)
3. Ciarkowska, K., Gambuś, F.: Micromorphometric characteristics of upper layers of soils contaminated by heavy metals in the vicinity of a zinc and lead ore plant. Polish J. Environ. Stud. **14**, 417–421 (2005)

4. Dovletyarova, E.A., Mosina, L.V., Petrovskaya, P.A.: Soil and environmental characteristics of the forest experimental station of RSAU of Moscow Agricultural Academy them. Bull. Peoples' Friendship Univ. Russ. Ser. «Agron. Anim. Ind.» **3**, 4–45 (2016). K.A. Timiryazev under plantations under conditions of different anthropogenic load
5. Dovletyarova, E.A., Mosina, L.V., Paltseva, A., Morin, T., Petrovskaya, P.A.: Soil-ecological characteristics of the recreational forest ecosystems in Moscow. Bull. Peoples' Friendship Univ. Russ. Ser. « Agron. Anim. Ind.» 4 (2016)
6. Grzybowski, W.: Comparison between stability constants of cadmium and lead complexes with humic substances of different molecular weight isolated from Baltic Sea water. Oceanologia **42**, 473–482 (2000)
7. Hutton, M.: Human health concerns of lead, mercury, cadmium and arsenic. Lead Mercur. Cadmium Arsen. Environ. **31**, 53–68 (1987)
8. Inboonchuay, T., Suddhiprakarn, A., Kheoruenromne, I., Anusontpornperm, S., Gilkes, R.J.: Amounts and associations of heavy metals in paddy soils of the Khorat Basin. Thai. Geoderma Reg. **7**, 120–131 (2016)
9. Kobierski, M., Dąbkowska-Naskręt, H.: Local background concentration of heavy metals in various soil types formed from glacial till of the Inowrocławska Plain. J. Elemntology. **17**(4), 559–586 (2012)
10. Kosheleva, N.E., Kasimov, N.S., Vlasov, D.V.: Factors of the accumulation of heavy metals and metalloids at geochemical barriers in urban soils. Eurasian Soil Sci. **28**(5), 476–492 (2015)
11. Mao, Q., Huang, G., Buyantuev, A., Wu, J., Luo, S., Ma, K.: Spatial heterogeneity of urban soils: the case of the Beijing metropolitan region. Chin. Ecol. Process. **3**, 1–11 (2014)
12. Martínez-Alcalá, I., Clemente, R., Bernal, M.P.: Metal availability and chemical properties in the rhizosphere of lupinus albus l. growing in a high-metal calcareous soil. Water Air Soil Pollut. **201**, 283–293 (2009)
13. Mosina, L.V., Dovletyarova, E.A., Andriyenko, T.N.: The Forest experimental station of Russian State Agrarian University - Moscow Timiryazev Agricultural Academy as object of environmental monitoring of forest and forest-park landscapes of the megalopolis of Moscow. PFUR, 221 (2014)
14. Mosina, L.V., Dovletyarova, E.A., Petrovskaya, P.A.: Microbiological assessment of a condition of forest and forest-park ecosystems. «Agron. Anim. Ind.». Bull. Peoples' Friendship Univ. Russ. Ser. **4**, 42–51 (2015)
15. Nicholls, A.M., Mal, T.K.: Effects of lead and copper exposure on growth of an invasive weed Lythrum salicaria L. (Purple Loosestrife). Ohio J. Sci. **103**, 129–133 (2003)
16. Saint-Laurent, D., Hähni, M., St-Laurent, J., Baril, F.: Comparative assessment of soil contamination by lead and heavy metals in riparian and agricultural areas (Southern Québec, Canada). Int. J. Environ. Res. Public Health. **7**, 3100–3114 (2010)
17. Ross, G.L., Ponirovskaya, Y.: Lead and Human Health : An Update, p. 80. Science (2000)
18. Węgrzyn, M., Wietrzyk, P., Lisowska, M., Klimek, B., Nicia, P.: What influences heavy metals accumulation in arctic lichen Cetrariella delisei in Svalbard? Polar Sci. **10**, 532–540 (2016)
19. Unit Weight: Physical Properties of Soils and Compaction, pp. 1–10 (2000)

Postindustrial Space: Integration of Green Infrastructure in the Center, Middle and Periphery of the City

E. Yu. Zaykova(✉)

Agrarian-Technological Institute, RUDN University,
Miklukho-Maklaya Str., 8/2, Moscow 117198, Russia
zaykova_eyu@pfur.ru, lena_landscape21@mail.ru

Abstract. Industrial areas occupy about 18,800 ha; more than 19% of Moscow [7]. These areas consist of former production zones and abandoned landscapes with numerous warehouses and dumping grounds. More than 4300 hectares of the industrial areas will be revitalized in the near future. The first stages of construction have already begun, with projects such as "Tushinsky Airfield" and "ZIL". Other projects that are planned for construction over the next three-five years include: "Altufyevskoye Shosse", "Silikatnye Ulitsy", "Ogorodny Proyezd", "Magistralnye Ulitsy", "Severyanin", "Yuzhny Port", and "Paveletskaya". These reconstructions provide an opportunity for sustainable development of new housing and social infrastructure with the restoration of natural ecosystems in the city. Designing in harmony with nature generates the potential for dynamic development of a resilient city [2]. This approach would alter the expansive and strictly-industrial areas into a "green and blue" city framework with natural water features and green areas. Incorporation of this spatial structure model would enhance city development at all levels; effectively stimulating the economy of the Moscow region. To achieve sustainable urbanization, it is necessary to use innovative restorative design; incorporating modern construction technologies while creating harmony and preserving heritage, to optimize forest areas, water areas, and open spaces, ultimately benefitting recreation, community-building, residential and office clusters [15].

Keywords: Sustainable urbanism · Resilient city · Post-industrial area · Biourbanism · Means of environment identity · Central · Middle and periphery of the city · Green and blue frame · Landscape urbanism

1 Introduction

Urban green infrastructure development, or "landscape urbanism", implies environmental and natural component restoration along with expansion of urban boundaries, such as incorporation of "abandoned" landscapes into the urban structure [11]. When creating a new territory development strategy, the natural environments of post-industrial reality and progressive urbanization must be carefully considered in the spatial development of cities. It is also necessary to reference new industrial area development models for new urban renovation ideas where the landscape typology and

V.I. Vasenev et al. (eds.), *Megacities 2050: Environmental Consequences of Urbanization*,
Springer Geography, https://doi.org/10.1007/978-3-319-70557-6_12

natural features have formed the environmental identity [16]. Moscow's rust belt is not structurally homogeneous. Industrial landscape typologies require individual design, and as such, can give the metropolis a new image of a "green city" [19]. Modern examples of this approach are the following projects: "Songzhuang Creative Clusters" in China, "The Delta District" in Vinge (Denmark), the new residential development on the periphery of Hamburg (Germany), "Arabianranta and Ruoholahti" in Helsinki (Finland), "Hammar by Sjostad" in Stockholm (Sweden), "Oerlilerpark", living quarters in Zurich (Switzerland), and many others. These examples demonstrate not only an integration of the former industrial areas and agricultural land into the new urban fabric, but also have the most significant impact on the landscape of the surrounding areas with transitional boundary balances between the natural and constructed, urban structures and a person's living space. The hybridization of strategies and urban industrial developments in these examples are used as the basis for a comprehensive study and urban policy adaptation into Russian city frameworks.

2 Goals and Objectives

The goal of this study, is to present the former industrial landscapes of Moscow in the light of the landscape urbanism. The research objective is to conduct a comprehensive assessment of the impacts of the sustainable redevelopment of these areas on surrounding areas, and of subsequent development of combined strategies for urban planning at different planning levels.

3 Materials and Methods

The distribution of the industrial areas in the central part of the city is a result of the constant historic center built-out. As old constructions were rebuilt, surrounding landscapes were not considered (Fig. 1).

Fig. 1. Location of industrial areas in the middle and peripheral parts of Moscow [7]

In the first redevelopment projects, "Krasny Oktyabr" and "Danilovskaya Manufactura" for example, the shortage of natural spaces were partially offset by the proximity of the Moscow River acting as the main central urban feature [9]. Green space in the central part of the city can be incorporated by using new technologies to vegetate the horizontal and vertical surfaces of buildings, and by connecting preserved natural areas through landscape design [10, 12]. Examples of transformations of different city surface types are presented in "Songzhuang Creative Clusters"; a project in China in which the urban fabric is integration with open spaces and agricultural landscapes through the creation of vertical gardens, rooftop gardens, and gardens with rainwater collection systems [17]. This approach increases the environmental sustainability of a particular place, providing aesthetic transitions through the residential and commercial territories by unifying the separate areas of the urban fabric through landscape design. This approach would overcome the shortage of natural spaces in the central part of Moscow and provide similar benefits to the urban landscapes and environment. Strategies for improving quality of the urban structure are shown defined in Table 1 and described below.

Table 1. Strategies for improving urban planning in the urban centers

Economic feasibility (attractiveness)	**" Flowing " surfaces**	**Facade transformation**	**False-frames integration**

Stuttgart-Münster [photo by author]	Copenhagen – Sloping roofs in residential areas [photo by V.A.Nefedov]	"Green" roof of the Sihlcity shopping center, Zurich [photo by author]	False facade in the loop of office building – Zurich [photo by author]
Increasing green areas in the building structure up to 50-80 percent serves as payback costs, improved company image, as well as the social attractiveness of the facility for different activities	The inclined surfaces in the central part of the city provide "green" architecture, tying industrial building design elements with surrounding area by means of green surfaces	Use of vertical and horizontal surfaces with for "green" design, connects the city's grey elements through green infrastructure	Facades used in reconstruction and restoration of historical and cultural heritage buildings, ensure the "life" of historic elements in the new socio-economic and environmental conditions

Economic feasibility (attractiveness) strategies are used to increase the green surfaces in buildings' structure with the objective to cover expenditures, improve a company's image, as well as create social attractiveness of industrial areas for a variety of activities.

"Flowing" surfaces strategies use materials and technologies that impart modern designs of the trending industrial rehabilitation. This strategy is particularly applicable to the re-purposed central-city industrial areas; connecting the inner and outer industrial design areas through "green" landscaped surfaces.

Facade transformation strategies take advantage of the vertical and horizontal surfaces, by vegetating them and connecting them other elements of the green framework and green infrastructure of the city. The inadequacy of this strategy is the absence of rooftop gardening on medium and low-rise buildings that are preserved in city's historic centers. The basic requirement of the green technology is to connect the greened surfaces with the surrounding architectural landscape through either the reconstruction of old buildings or integration of new building designs. The historical areas, operational feasibility, and modern engineering suitability must all be taken into account. The implementation of roof-gardens would compensate for the lack of green spaces in built-up historical centers. The Scandinavian projects illustrate that these green-roofs increase a city's green space, improve roof life in harsh climate areas, have multi-seasonal aesthetic appeal, leading to an overall improved city quality of life (Fig. 2).

Fig. 2. Underground parking roof with low-maintenance plants in the Hammar by Sjostad residential area of Stockholm, Sweden [photo by author].

False-frames integration strategy is used for the reconstruction and restoration, particularly of historical and cultural heritage constructions to ensure the endurance of old architecture and buildings in new socio-economic and environmental conditions. These technologies are extremely important for renovating industrial areas in the central historical district, and also for providing new surface design in industrial architecture. For example, the dwelling in Konstanz was redesigned with green-walls, uniting the building with the environment. Such methods are applicable to harsh climate countries, like Russia (Fig. 3).

Fig. 3. The dwelling reconstruction with the false facade for vertical gardening - Konstanz, Germany [photo by author].

These strategies and technologies provide the urban built-up environment with aesthetic elements, a sense of community and enable the city center to function sustainably [4].

The industrial areas are densely distributed in the central district of Moscow and less so toward the periphery. Presently, 4300 hectares (4.4% in the periphery of the city) have been selected for reconstruction until 2025 [7]. These areas are made up of mixed land uses: dense building infrastructure, residential dwellings, and industries. They also have poor social and cultural infrastructure. The green spaces in the periphery of the city are made up of protected areas, national parks, forests, and public gardens, totaling an area of 16,000 ha. These facts present a large opportunity for a new strategic approach for urban redevelopment in Moscow. The former industrial areas can be redeveloped using innovative approaches to integrate the surrounding natural spaces. Many European countries have implemented sustainable systems between nature and the built urban environment, and have demonstrated the wide possibilities for this change, leading to the optimization of social, environmental, functional, and aesthetic elements of industrial areas in different stages. When implementing the new project "Delta District" in Vinge city in Copenhagen, the many nuances of construction planning were taken into account, such as, environmental issues, the spatial organization of the infrastructure, and the multi-functionality of the urban spaces and features, to determine the identity of the environment and its different effects on different parts of the urban space [2]. The environment identity of the city is defined by the industrial areas, landscape typology, and land uses, and their relationships with the boundaries of unique spaces within the urban fabric, as well as the integration of these factors with the city's core structures. To create a new resilient city, with improved and safe environmental conditions for all residents, it is necessary to consider all of the industrial area parameters mentioned above to determine the best methods for their optimal integration.

Method of functional restructuring. The first step is to restructure the industrial areas in the middle and peripheral parts of the metropolis, connecting natural areas with wastelands and industrial architecture. The second step is to restructure areas in the middle circuit of industrial area by siting areas with potential for green infrastructure. Linear natural spaces in the landscaped plots of vehicle-pedestrian communications, parks, plazas, and waterfronts are green spaces that can be integrated into the industrial structure and into the city frame [4]. The main objectives are to connect the middle and peripheral industrial areas with the city center, and to create a convenient, pleasant, and safe system for pedestrian communication, including bicycle routes, pedestrian plazas, sidewalks, and other transit routes. The industrial area reconstruction in Zurich (Switzerland) – Oerlilerpark, and living quarters, Pfingstweidstrasse and Turbinenplatz, are examples where residential and business clusters are linked to the center of Zurich-West with vehicle and pedestrian arteries designed with green infrastructure [20] (Fig. 4).

Fig. 4. Former industrial space turned into a residential and business clusters with renewed nature in the structure of new territories - Zurich, Switzerland [photo by author].

Functional restructuring includes defining each functional landscape area on the periphery of the city and reserving suburban space for the industries moving out of the urban center. Implementing this restructuring strategy on the border of Moscow and the

Moscow region, requires development of social and environmental policies for the joint uses of natural resources for the sustainable development of the region. Outside of the city, in the suburban, industrial, and new built-up areas, landscape development and connectivity requires a combination of contemporary approaches to implement sustainable transport and business development [18].

Method of natural ecosystems regeneration. The current trend of sustainable urbanism, marked by Farr [5], demands that the transportation system be integrated with land-use and technologies in a specific ratio, along with the development of biodiversity corridors. Biourbanism as a paradigm of urban design and planning that could help in the development of sustainable territories of the cities' suburbs, including Biosystems and urban systems [13]. Some parameters of these systems are aimed at various objectives such as the preservation of the landscape, economical factors, collection of water resources etc. The structure of the territory and technology play an important role in sustainable environmental development. The conservation of natural habitats, spontaneous nature, and the nature of wastelands within industrial areas, all have great potential for adaptation into the new functions of urban development [3]. Other examples of opportunities are: the restructuring of industrial areas in the peripheral water areas of the city to optimize aquatic vegetation, the optimizing of rainwater harvesting for the balance of the ecosystem and environment as a whole. The "green connection" of Schanzengraben, and the rainwater collecting system supplying the surrounding landscape in Turbinenplatz in Zurich are some European examples of ecosystems-regeneration of former industrial and other city areas. In the first example, a former military fortress was transformed into an attractive space using green technologies, influencing the development of all accessible areas of the city (Fig. 5).

Fig. 5. The transformation of the uncomfortable spaces to leisure area with the stored and integrated nature - Zurich, Switzerland [photo by the author].

Schanzengraben is a linear park by the water with a revitalized natural environment that connects the former industrial area of the district center Turbinenplatz with the Zurich-West center. Regenerating and adapting industrial areas safeguards the natural areas. The reconstruction of a turbine factory in one of the largest post-industrial areas in Zurich is a successful demonstration of the possibility of creating such rainwater collection and redistribution technologies (Fig. 6).

Fig. 6. Turbinenplatz: the channel for rainwater collection and the natural landscape supply - Zurich, Switzerland [photo by author].

The main concept of the design involves the industrial history of the place. Large areas of hard surfaces with a system of storm-water outlets, and the redevelopment of buildings in the business and office clusters offer the possibility of capture and redistribution of a large amount of rainwater to the landscapes of the vertical gardens of the surrounding buildings [20].

Method of water communications development. The Moscow River is the main feature of the central part of the city. The waterbody system in the middle in the periphery of the metropolis is connected with a number of industrial zones. This makes cultural and economic revival of such spaces possible [8]. One of the important elements of revitalizing the river is connecting the development of communications systems to the water and on the water. Types of leisure areas and near-water traffic are crucial factors for pedestrian transport system reorientation, in ensuring safe accessibility of the river for all different types of users. In the case of industrial areas, typological structure integration with new hybrid space is capable of transforming the ecosystem of the area as a whole. Global practice of redesigning post-industrial areas in these waterbody areas offers many examples of mutual integration of industrial areas with the use of new technologies for artificial channels development for improved environment comfort and an increase in biodiversity. As an example, the Hammar by Sjoestad residential area in Stockholm is a public space that was organized around a structural framework of the piers, with walkways along the shoreline and viewpoints of the surrounding landscape, providing a variety of contact with water [1] (Fig. 7).

Fig. 7. Hammar by Sjoestad - water areas loop near housing with communication to the water, along the water and on the water - Stockholm, Sweden [photo by author].

The introduction of new construction technology enables development of methods for transformation of river ecosystems in order to enhance the river's long-term protection and sustainability. The development of artificial channels in industrial areas, connects the waterbody system, creating natural components, improving the water and aquatic vegetation, and effecting the environmental change, altogether creating an ecology identity for a particular place [14].

Method of nature identification. Identity is a very important quality that can lead to the reconstruction of natural spaces and the return of nature in our cities. Conservation and adaptation of urban industrial landscapes involves following the principles of ecological design and considering the natural habitat with a deep knowledge and understanding of the dynamics of trees, shrubs, perennials and meadow plants. Landscape analysis of an industrial space and a study of topography is performed to inform the preservation of the natural areas. Species composition and vegetative cover of natural habitats are considered and are very important when integrating natural systems. They can either naturally complement a natural oasis, or create a landscape composition on the "principle of contrast". This depends on the overall design concept and typology of the landscape. Identification of the natural habitat in the urban environment is demonstrated in the Park am Nordbahnhof project in Berlin. According to the authors of landscape architectural Bureau Fugmass Janotta and architectural firms Atelier Loegler, it is the result of a step-by-step development of a project idea that consulted environmental objectives and requirements for intensive-use recreation resulting in what they called "creative compensation" [6].

Renovations of the industrial areas of Moscow must consider the possibility of a gradual withdrawal of industry from the city borders. The government is also considering requiring some industries to focus on cleaner production or promote scientific development in their territory. Thus, siting potential natural areas and implementing green infrastructure, such as vegetated roof and walls of buildings and facilities, becomes important in the regeneration of industrial areas. Therefore, the reconstruction of existing industries encompasses a broad range of issues: from the creation of not only "green" infrastructure and an improved working environment, but also the sustainable design and landscape of industrialization. An example of such an approach to landscape organization of an existing industrial site in the city is given by the Münster power plant station in Germany (Fig. 8).

Fig. 8. Münster power station - operating industrial facility with innovative technology, green design - Stuttgart-Münster, Germany [photo by author].

This renovation changed the image of industrial production to one with a focus on "environment construction". It was a creation of meaningful integration of green components in the structure of an industrial building. With existing industrial areas incorporating new green technology, and the sprawling urban environment and expanding living space incorporating an increasing number of gardens and green spaces, the noise pollution and environmental stress at these sites is reduced.

4 Results and Discussion

The results of the analyses of the industrial areas in different parts of the Moscow, have demonstrated a great potential for new landscape, ecological, recreational and social design. The complexity of the structure and diversity of the industrial landscape typology presents an opportunity to shape the image of a "green city" while maintaining natural systems. However, the analysis of the central part of the city has revealed the lack of historical architecture connections with the natural environment, due to lack of space. However, new technologies and strategies now exist that integrate green spaces by taking advantage of the horizontal and vertical surfaces of buildings. Successful modification of industrial areas landscape is guaranteed by a competent restructuring of territories and adaptation to spontaneous vegetative areas, wastelands' natural landscapes, old industrial architecture, and the main elements forming the urban fabric.

5 Conclusion and Recommendations

Modifying urbanization strategies appears to be more possible now than it was in the past, as appreciation for the connection between urban spaces and nature grows. The solution to the problem lies in the search for new strategies of landscape reform, adaptation, and maintenance of the retrofitted spaces in the context of functional restructuring, regenerating natural ecosystems, and identifying and transforming building facades and frames. By combining these methods in the industrial area renovation process, we can attain an environmentally sustainable and socially acceptable space in harmony with the landscape typology in different parts of the city. The proper application of "nature" is one of the most important elements of successful and long-term sustainable renovations of industrial.

References

1. Andersson, Th.: Utopia in Green and Blue. TOPOS: The International Review of Landscape Architecture and Urban Design, No. 69: Reuse, pp. 64–70 (2009)
2. Andersson, S.L.: The Urban as a resilient system of processes. The Delta District, Vinge, and Denmark. TOPOS: The International Review of Landscape Architecture and Urban Design, No. 90: Resilient Cities and Landscapes, pp. 22–29 (2015)
3. Arabianranta: Rethinking Urban Living. (City of Helsinki Urban Facts. City of Helsinki Economic and Planning Center. Art and Design City Helsinki Oy). WS Bookwell Oy, Porvoo, p. 288 (2007)
4. Elsaesser, Ch: New Living in Jenfeld. TOPOS: The International Review of Landscape Architecture and Urban Design, No. 70: Sustainability, pp. 36–41 (2010)
5. Farr, D.: Sustainable Urbanism: Urban Design With Nature, 352 p. Wiley, Hoboken (2008)
6. Hentschel, A.Y.: How green will we live. Gruener Wohnen/Green living [Text]/Deutscher Landschaftsarchitektur-Preis/German Landscape Architecture Prize, pp. 54–61 (2011)
7. Information on StroiMos. www.stroi.mos.ru
8. Information on ArhsovetMsk. http://archsovet.msk.ru/competitions/moskva-reka/meganom-moskva-reka
9. Information on. http://www.dm1867.ru/quarter.php/
10. Information on. http://www.zinco.ru/
11. Krasilnikova, E.: Landscape Urbanism. Teoriya-Praktika: scientific monograph/LLC IAA Oblastnyevesti, Volgograd, p. 1. Scientific and practical bases of landscape urbanism, p. 156 (2015). (in Russian)
12. Lysikov, A.B.: Vertical gardening: design decisions. Choice of plants. CJSC Fiton +, 136 p. (2012). (in Russian)
13. McGregor, A.: The Biourbanism Paradigm. TOPOS: The International Review of Landscape Architecture and Urban Design, No. 90: Resilient Cities and Landscapes, pp. 90–95 (2015)
14. Neighbours Ruoholahti: 1992–2002 the first 10 years people – jobs – townscape, p. 208. ISBN 952-99059-0-4
15. Nannan, D., Zhang, L., Ruff, S.: From Expo City to Sustainable City. TOPOS: The International Review of Landscape Architecture and Urban Design, No. 70: Sustainability, pp. 19–27 (2010)
16. Nefedov, V.: How to return the city back to people. Iskusstvo – XXI век, 160 p.: ill (2015). (in Russian)

17. Pieprz, D., Grove, M.: Songzhuang Creative Clusters. TOPOS: The International Review of Landscape Architecture and Urban Design, No. 84: Urban Strategies, pp. 44–50 (2013)
18. Rolandsen, S.E., Dirckinck-Holmfeld, K.: Suburbs of the Future: Lessons Learned. TOPOS: The International Review of Landscape Architecture and Urban Design, No. 84: Urban Strategies, pp. 64–71 (2013)
19. Uhrig, N.: Reset at Gleisdreieck in Berlin. From an Inner-city Wilderness to a Metropolitan Park. TOPOS: The International Review of Landscape Architecture and Urban Design, No. 85: Open Space, pp. 14–21 (2013)
20. Wullschleger, P., Oezkan, H., Handschin, L.: Landscape Architecture in Zurich – A Historical Timeline. TOPOS: The International Review of Landscape Architercture and Urban Design, No. 73: Sustainability, pp. 1–14 (2010)

Vegetation Indicators of Transformation in the Urban Forest Ecosystems of "Kuzminki-Lyublino" Park

A. Yu. Buyvolova[1(✉)], T.A. Trifonova[2], and E.P. Bykova[2]

[1] Peoples' Friendship University of Russia, ATI, Moscow, Russia
buyvolova@gmail.com
[2] Faculty of Soil Science, Lomonosov Moscow State University, Moscow, Russia

Abstract. Urban forests usually cover slightly disturbed or natural soils. The morphological structure and dominating species in urban forests are usually like the natural vegetation. These similarities constrain distinguishing between an urban and rural forest without detailed analysis of soil and vegetation properties. Urban forests are exposed to negative city influences, of which air pollution is one of the most substantial. The continuous anthropogenic influences alter the structure of urban forest. This article presents vegetation indicators of changes that occurs in the structure of the forest ecosystems as the response to an anthropogenic impact. The following indicators of forest ecosystem transformation in urban conditions were investigated: (i) reduction of the projective cover of moss layer, until its complete absence (in the pine forest), (ii) increasing the activity of adventive species (*Acer negundo* L.) in the undergrowth, (iii) high variability of the phyto mass of the ground herbaceous vegetation, and (iv) change in the arial distribution of adventive species of the ground herbaceous vegetation.

Keywords: Urban forest · Vegetation indicators · Bio indication · Protected area

1 Introduction

Forest ecosystems in a city perform multiple functions as a part of urban developmental planning and at the same time as a component of natural environment. These functions include environmental benefits (e.g. ecosystem formation o, air filtration, noise pollution reducing, etc.) and social (e.g. recreational, educational). The importance of the urban s forest' functions requires for the special attitude to their management and study. The management of an urban forest in a mega polis has to combine natural conservation efforts with maximizing the accessibility of the citizens to the recreation resources. In terms of management and decision-making, this task is very challenging.

In Moscow, natural city forest ecosystems have been preserved in protected natural areas. These protected areas are the basis for the biological regulation and stabilization of the urban environment. The necessity for the assessment and conservation of natural

V.I. Vasenev et al. (eds.), *Megacities 2050: Environmental Consequences of Urbanization*, Springer Geography, https://doi.org/10.1007/978-3-319-70557-6_13

areas in the cities has been confirmed by the studies of researchers as well as by legislation and strategic documents of development for metropolises all over the world [2, 12, 13, 15].

2 Materials and Methods

We analyzed the transformational indicators of natural ecosystems in the 'green belt' of the city of Moscow by comparing them with similar ecosystems located outside the city. Such reference ecosystems were selected at the territory of the forests of the Prioksko-Terrasny biosphere reserve (BR). The Prioksko-Terrasny BR locates in the south of the Moscow region and its macro-climate conditions corresponds to the climatic characteristics of Moscow [8, 9]. The Prioksko-Terrasny BR and the "Kuzminki-Lyublino" Park both locate at the terraces above the flood-plains of large rivers (Oka and Moscow) in the area dominated by Albeluvisols on glacio fluvial sandy deposits. Since the climate and lithological conditions at the research and reference sites are similar, comparing the structural indicators of vegetation between the sites is relevant.

When selecting and allocating the permanent sample plots, typicality and representative sites were chosen. The assessment of the city impact on the vegetation in the "Kuzminki-Lyublino" Park was conducted at two key plots, which were the least affected by anthropogenic impacts. The key plots differed in plant communities: complex pine dominated the first plot, whereas the birch forests covered the other plot. The natural counterparts with similar forest types were selected in the Prioksko-Terrasny BR. The pine forests were selected as model plots, since a pine (*Pinus Sylvestris* L.), is a widespread speciment. Besides, it is a sensitive indicator of human influence, responding to the impact by defoliation and it is resistant to fairly high levels of pollution [5–7]. Birch forests, (*Betula pendula* L.) have been selected as secondary growth forests, during the stage of recovery after being felled. Pine and birch forests are the most widespread forests in the Moscow Region. In addition, the selected forests are most common to the area in the "Kuzminki-Lyublino" Park and in the BR.

Information about the age of the stand of trees has been derived from the diameter and height of the tree stands based on the forest survey data. For the diagnosis of the transformation of urban forests, the botanical descriptions of study plots were made within the key areas with a five-fold repetition. The names of forest types in the sample plots are given according to the typological classification of groups of forest types in the Moscow region [12]. The projective cover is determined visually and expressed as a percentage, an abundance – on the Brown-Blanket scale of plant abundance. Species diversity of plant communities has been studied with the help of isolating the alpha and beta diversity [9]. Ecological-coenotic groups (ECG) were determined based on the relationship of species to individual environmental factors [3, 4]. The synantropysation was calculated according to the percentage of the adventive species in the cover [1].

3 Results and Discussions

To evaluate the environmental performance of the permanent sample plots, geobotanical descriptions were processed using H. Ellenberg's ecological scale, evaluating the characteristics of soil moistening, shade density-suppression, nitrogen level, the availability of mineral elements in soil, soil acidity [4]. The average scores of the habitat's environmental parameters indicate differences between the plots (Table 1).

Table 1. Characteristics of the habitats of "Kuzminki-Lyublino" park and the Priosko-Terrasny BR

Forest type	Territory	t°	Soil humidity	Soil acidity	Nitrogen level.	Light availability
Pinewood	Urban forest	5.3	4.8	4.8	4.9	4.8
	Nature reserve	5.3	4.4	3.9	3.9	4.9
Birchwood	Urban forest	5.1	4.3	5.3	4.0	6.6
	Nature reserve	5.3	4.6	6.3	4.1	6.1

The leading environmental factor contributing to higher species richness in the birch forest of "Kuzminki-Lyublino" Historical Park differed from those in the complex pine forest in the nature reserve. The main reason was likely the better shade density in the PR. Despite the fact that in some cases the geobotanical descriptions for an urban forest park and nature reserve are quite varied, we can conclude, based upon Ellenberg's scale that the vegetation of phytocenoses selected for the study of both protected areas can be considered as quite aligned in relation to the environmental factors.

The characteristics of the forest stand of birch forest of studied territories were compared. The crown density determined the main difference in the forest stand of the studied areas (Tables 2 and 3). In "Kuzminki-Lyublino" Natural-Historical Park these indicators are two times lower. The species-edificators and dominants have been used to characterize and to identify the park's forest ecosystems. They are the builders of plant communities, and largely determine their composition, structure, regeneration and metabolic processes [14].

Table 2. Comperative assessment of birchwood stand in studied territories

Territory	Stand composition	Height, m	Age, years	Crown density, %
Urban forest	Birch - 90%; Pine - 10%	25	40–60	5–30
Nature forest	Birch - 80%; Aspen - 10%; Pine - 10%	25	50–70	30–40

Table 3. Comperative assessment of Birchwood underwood in studied territories

Territory	Underwood (density, height)	Underbirch (density)	Plant cover	Plant cover dominant	Total species number (advent.)
Urban forest	Acer + Populus (rarely, 1 m)	Sorbus, Frangula, Corylus. (rarely)	70–80, %	*Calamagrostisepigeios, Convalariamajalis, Agrostiscapillaris*	73 (1)
Nature forest	Quercus + Populus + Betula (rarely, 1 m)	Euonymus, Genista (rarely)	70–80, %	*Carexpilosa, Calamagrostisarundinaceae, Calamagrostisepigeios*	54 (0)

A comparative analysis of the birch forests of the Prioksko-Terrasny BR and the "Kuzminki-Lyublino" Park showed the following main differences:

1. The birch forest in the "Kuzminki-Lyublino" Park was represented by even-aged forest stands, while the birch forest in Prioksko-Terrasny BR an uneven-aged stand was described;
2. In similar indicators of stand normality, the crown density differed substantially. Through the crown sparsity the presence of light-demanding herb stratum in the sample plots "Kuzminki-Lyublino" Park was defined. One adventive species occured – *Impatiens parviflora L.*;
3. The absence of an adventive species in the birch forest of the Prioksko-Terrasny BR, and single representatives in the "Kuzminki-Lyublino" Park was noted. The specific diversity in the urban forest was significantly higher than that of the Prioksko-Terrasny BR.

A comparative analysis of the complex pine forests (Tables 4 and 5) of the park and the nature reserve showed the following main differences:

1. If in the pine forest of the nature reserve a moss layer covered 90%, whereas it was completely absent in the "Kuzminki-Lyublino" Park;
2. In the advance growth of the "Kuzminki-Lyublino" Park one an adventive species, (*Acer negundo* L.) was widely spread;

Table 4. Comperative assessment of pinewood stand in studied territories

Territory	Stand composition	Height, m	Age, years	Crown density, %
Urban forest	Pine 70% Maple 20% Birch 10% + Spruce	30	100	40–60
Nature forest	Pine 19% + Birch < 10%	30	100	50–60

Table 5. Comperative assessment of pinewood underwood in studied territories

Territory	Underwood (density, height)	Underbirch (density)	Plant cover	Plant cover dominant	Total species number (advent.)	Moss cov., %
Urban forest	Acer platanoides + Quercusrobur + Acer negundo (up to (30%), 7 m)	Sorbus, Euonymus, Sambucus (10%)	<5%	Convalariamajalis, Luzulapilosa, Impatiens parviflora	34 (1)	0
Nature forest	Quercusrobur + Betulapendula(up to 20%), 10 m)	Sorbus, Frangula, Sambucus, (10%)	7–25%	Convalariamajalis, Vacciniumvitis-idaea, Calamagrostisarundinaceae	28 (1)	90

Alpha-diversity within the community was defined by the following parameters: species wealth – the total number of species (TNS) per unit area (number of species/10 m^2) and species saturation (Table 6).

Table 6. Biodiversity of studied territories.

Biotop	Territory	Species saturation	TNS	Jaccard index	Whittaker index
Birchwood	Urban forest	39.2 ± 2.1	73	0.24	0.86
	Nature forest	23.6 ± 2.7	54		1.29
Pinewood	Urban forest	20.4 ± 1.5	34	0.24	0.67
	Nature forest	15.8 ± 1.1	28		0.77

Beta-diversity shows the degree of differentiation of the distribution of species between communities. The highest species saturation (73 species/ha) was obtained for the birch forest of "Kuzminki-Lyublino" Park, since here the meadow marginal vegetation dominated. This indicator was also sufficiently high in the birch forest of the Prioksko-Terrasny BR (54 species/ha). The data obtained correlated well with the average values of soil wealth and shade density factors, which fully complied with *ECG* communities. For the pine forest of "Kuzminki-Lyublino" Park the species saturation (34 species/ha) was also higher than that of the Prioksko-Terrasny BR (28 species/ha). The prevalence of species in the city park was insignificant. The loss of the vegetational biodiversity cover was not evident.

The values of the Whittaker index for vegetation communities studied in both the studied area show that the floristic heterogeneity was slightly higher in an urban park than in the nature reserve, but this was not due to adventive species. The study of the synanthropization level has not revealed an increase in synanthropic species. The vegetation of the "Kuzminki-Lyublino" Park green belt relates to the natural communities.

The similarity of communities in the nature reserve and the park were also assessed through the use of Jaccard index. The analysis showed that the similarity coefficients for both plant communities were equal 0.24 (Table 6). Table 7 is a comparison of the bio-productivity in the herbaceous-suffruticous layer of the birch forest of the "Kuzminki-Lyublino" Park and the reference territory.

Table 7. A comparison of the bio-productivity in the herbaceous-suffruticous layer of the birch forest of the "Kuzminki-Lyublino" Park and the reference territory.

	Moisture content, 100%					Moisture content, 0%				
	Urban forest		Nature forest		t-test	Urban forest		Nature forest		t-test
Year	m, kg/ha	V, %	m, kg/ha	V, %		m, kg/ha	V, %	m, kg/ha	V, %	
2014	290.2	24.0	350.1	14.0	<2	110.4	16.0	140.3	14.0	<2
2013	470.0	39.0	540.9	6.0	<2	130.3	15.0	180.0	7.0	<2

Plant biomass characteristics corresponded to zonal. The features for above-ground plant biomass in the herbaceous-suffruticous layer for 2014 and 2013 showed higher than average values of plant biomass in the Prioksko-Terrasny BR due to presence of sedge grass, however, statistically significant values were not found. It has been shown that the coefficient of variation values (CV) in the city was always higher than in the nature reserve. The same results have been shown by the values of the projective cover for the herbaceous tier. Thus, the projective cover and bio-productivity indicators of the studied have areas differed, however the variation of vegetation in urban forest was high.

4 Conclusion

The vegetation cover is one of the most dynamic components of the ecosystem. In the city environment it is subjected to structural changes. The following vegetation indicators were approved as relevant to monitor the transformation of natural forest ecosystems in the city of Moscow: a reduction or complete absence of the of moss layer (in the pinewood), increasing activity of adventive species (e.g. *Acer negundo* L.) in the undergrowth, high variability of the phyto mass in the ground herbaceous vegetation, and a change in arial distribution of adventive species.

Further researches with a view to identifying indicators of vegetation transformation in the city of Moscow are needed. It will allow to notice reversible changes in the early stages in the transformation of urban ecosystems. The development of management technologies of the urban forest can appeal the proposed indicators.

Acknowledgements. The publication was financially supported by the Ministry of Education and Science of the Russian Federation (the Agreement # 02.A03.21.0008) and by Russian Foundation for Basic Research #15-04-04702, RF.

References

1. Abramova, L.M.: Synantropization of vegetation: patterns and control of the process (case study of Republic of Bashkortostan): abstract of dissertation. Perm State University, 46 p. (2004) (in Rusian)

2. Carreiro, M.M.: The growth of cities and urban forestry. In: Carreiro, M.M., Song, Y.-C., Wu, J. (eds.) Ecology, Planning and Management of Urban Forests: International Perspectives. Springer, New York (2008)
3. Check-list of the vascular flora of the Central Russia. http://www.jcbi.ru/eco1/index.shtml. Accessed 01 Sept 2015
4. Ellenberg, H., Weber, H.E., Dull, R., Wirth, V., Werner, W., Paulissen, D.: Zeigerwerte von Pflanzen in Mitteleuropa. ScriptaGeobotanica **18**, 1–248 (1991)
5. ICP IM Manual for integrated monitoring., Compiled by the ICP IM Programme Centre Finnish Environment Institute, Helsinki, Finland (2010). http://www.syke.fi/en-US/Research__Development/Maintaining_ecosystem_services_and_biodiversity/Monitoring/Integrated_Monitoring/Manual_for_Integrated_Monitoring
6. Klánová, J., Čupr, P., Baráková, D., Šeda, Z., Anděl, P., Holoubek, I.: Can pine needles indicate trends in the air pollution levels at remote sites? Environ. Pollut. **157**(12), 3248–3254 (2009). https://doi.org/10.1016/j.envpol.2009.05.030
7. Lamppu, J., Huttunen, S.: Relations between scots pine needle element concentrations and decreased needle longevity along pollution gradients. Environ. Pollut. **122**, 119–126 (2003). https://doi.org/10.1016/S0269-7491(02)00274-9
8. Lokoshchenko, M.A., Korneva, I.A.: Underground urban heat island below moscow city. Urban Clim. **13**, 1–13 (2015). https://doi.org/10.1016/j.uclim.2015.04.002
9. Magurran, A.E.: Measuring biological diversity. Blackwell Publishing, Oxford (2004). 256 p.
10. Pchelkin, A.V., Pchelkina, T.A.: Influence of polluting substances and climatic anomalies on fauna of reptiles and amphibious in the Moscow region The state and environmental pollution review in the Russian Federation for 2011. [Obzorsostoyaniyaizagryazneniyaokruzhayushcheisredy v RossiiskoiFederatsiiza 2011 g. Moscow, pp. 190–192 (2012)
11. Emery, M.: Promoting Nature in Cities and Towns: Practical Guide. Croom Helm, London (1986). 396 p.
12. Resolution of the Government of Moscow from 10 July 2014, N 394-PP, About basic provisions of new environmental policy of the city of Moscow for the period till 2030 (in Rusian)
13. Resolution of the Government of Moscow from 10 09 2002, N 743-ПП Regional Rules for management of urban green areas of Moscow (in Rusian)
14. Rysin, L.P., Savel'eva, L.I.: SosnovyelesaRossii [Pine Forests of Russia]. Moscow, 289 p. (2008) (in Rusian)
15. United Nations. The millennium development goals report 2008. New York, United Nations (2008). http://mdgs.un.org/unsd/mdg/Resources/Static/Products/Progress2008/MDG_Report_2008_En.pdf. (2016)

Environmental Monitoring of Sod-Podzolic Soils Under the Forest Stands over One Hundred Year Period: The Case Study at the Forest Experimental Station in Moscow, Russia

Lyudmila Vladimirovna Mosina[2],
Elvira Anvarbekovna Dovletyarova[1], Anna Paltseva[3,4,5](✉),
Tatiana Morin[3,5], and Polina Alexandrovna Petrovskaya[1]

[1] Department of Landscape Design and Sustainable Ecosystems, RUDN University, Miklukho-Maklaya Str., 8/2, 117198 Moscow, Russia
{elvira_2779, kinet_ppa}@mail.ru
[2] Russian State Agrarian University-Moscow Agricultural Academy, Timiryazevskaya Str., 49, 127550 Moscow, Russia
[3] Department of Earth and Environmental Sciences, Brooklyn College of the City University of New York, 2900 Bedford Avenue, Brooklyn, NY 11210, USA
apaltseva@gradcenter.cuny.edu, tatiana@usi.nyc
[4] Earth and Environmental Sciences, Graduate Center of the City University of New York, 365 5th Avenue, New York, NY 11016, USA
[5] New York City Urban Soils Institute, 2900 Bedford Avenue, Brooklyn, NY 11210, USA

Abstract. This study of forest soil ecology conducted in the forest experimental station of the Russian State Agrarian University of Moscow Agricultural Academy reveals the evolution of a soil profile over a 100-year period influenced by a change from pine and spruce stands to two-tier mixed plantations of birch, maple, linden and oak. The rates of formation of the humus horizon and the reduction of the thickness of the podzolic horizon was determined by this change. Native pine tree growth enhances the podzol development process resulting in the development of an admixture of deciduous tree populations. This growth of hardwoods succeeding into the pine forest turn reduces the rate of accumulation of the upper horizon by 3 times. Humus horizon under these mixed stands was found to be the thickest in the locations, hardwood dominated. Comparison of the taxonomic properties with the soil characteristics will help in developing criteria for the vegetation stability and productivity in urban and suburban areas. This study reveals that the composition of tree species and their successions influences the main soil forming processes in the profile of sod-podzolic soil. The results of the study can be used to predict the soil-forming processes and the states of the forests. The ability to assess and predict these conditions is extremely important for soil-ecological monitoring in recreational forests for ecological landscape mapping.

V.I. Vasenev et al. (eds.), *Megacities 2050: Environmental Consequences of Urbanization*, Springer Geography, https://doi.org/10.1007/978-3-319-70557-6_14

Keywords: Ecological monitoring · Soil evolution · Forest and forest-park landscapes · Unregulated recreation · Morphological profile of soils

1 Introduction

Environmental pollution has become a global environmental problem, having a negative impact on human health and jeopardizing our forests [1, 7]. In the light of the global environmental problems, environmental monitoring has become vital [10]. Forest and forest-park plantations are the key elements of such monitoring as they play a vital role in the health conditions of urban environments and their recreational loads.

In understanding the important role of the "green lungs" of our planet, their close interrelationship with soil quality must be recognized. Soil is a determining factor for the growth of productive and sustainable phytocenoses [4].

Forests gain significance when their unique compositions/arrays are contemplated; appreciated as monuments of nature, having great historical, scientific and practical value, and as systems delivering critical information on the long-term conditions of ecosystems.

Existing literature lacks data on long-term evolution of soil cover in the forest and forest park landscapes [5]. For this reason, the dynamics of such landscape changes were studied in the sod-podzolic soils (Eutric Podzoluvisols) under forest phytocenoses in Moscow for over one hundred years. This soil data (from the beginning of the last century; 1909–1910) was compared with the present-day conditions studied in this area. The obtained results provide information on the dynamic changes in soil-forming processes, helping to evaluate the evolution of the soil due to the influence of a variety of tree species and environmental factors, and to assess the state of the forest plantations and landscapes.

This study was conducted at the forest experimental station (FES) (55°50′ N and 37°14′ E) of the Russian State Agrarian University of Moscow Agricultural Academy (RSAU-MAAT). This unique forest occupies 232.6 ha of land [3] in the north-west of Moscow. The research has been being conducted at the FES for over 100 years [5]. The FES is impacted by unregulated recreation and nearby gas station emissions, the effects of which impose a certain anthropogenic load and imprint on the soil conditions and subsequently, the state of vegetation [4]. The research site consists of a mixed pine and spruce tree plot (quadrant 4, plot "Щ"), a native pine tree plot (quadrant 4, plot "E"), at two-tier plot with pine and birch in the first tier, and of linden and oak (quadrant 4, plot "Л") in the second tier.

2 Ecological Characteristics of the Plot "Щ", Quadrant 4

The plot "Щ" of quadrant 4 was established in 1891 by Dr. M.K. Turski in the north-eastern part of the FES with a compositional ratio of 6 spruce and 4 pine trees, to study the growth of mixed spruce-pine plantations [6]. The silvicultural characterization of the plot is presented in Table 1. The composition of this plot was maintained (the spruce 6: 1 pine ratio) until 1941 accounting for 1076 trees [5]. Later, severe

Table 1. Silvicultural characterization of the plot "Щ" (quadrant 4) of the FES

Year of recount	Age, years	Tree species	The number of trees, units/ha	Average		Compo-sition	Overall productivity, m^3/ha
				Diameter, cm	Height, m		
1891	25	Spruce	5409	5.5		6S4P	undefined
		Pine	839	11.5			
		Birch	38	7.9			
1898	32	Spruce	4589	13.5		6S4P	— » —
		Pine	738	8.9			
1904	38	Spruce	4239	16.7		6S4P	— » —
		Pine	732	11.2			
1909	43	Spruce	2650	18.2	12.5	6S4P	— » —
		Pine	660	12.1	17.5		
1914	48	Spruce	2353	20.2	14.0	6S4P	— » —
		Pine	547	13.6	20.5		
1923	57	Spruce	1859	22.0	15.0	6S4P	— » —
		Pine	484	15.0	21.0		
1930	64	Spruce	1502	24.3	17.0	6S4P	— » —
		Pine	439	17.5	22.5		
1935	69	Spruce	1076	26.1	19.0	6S4P	— » —
		Pine	382	20.8	23.0		
1941	75	Spruce	83	20.8	19.0	9P1S	542.9
		Pine	331	26.7	24.0		
1946	80	Spruce	13	23.2	19.0	9P1S	546.5
		Pine	299	26.4	24.0		
1951	85	Spruce	6	26.0	23.0	10P	576.3
		Pine	286	28.1	24.0		
1956	90	Spruce	6	26.0	23.0	10P	588.6
		Pine	229	28.6	24.0		
1961	95	Spruce	6	26.0	23.0	10P	588.6
		Pine	191	29.2	24.0		

drought removed spruce phytocenosis and the pine stands became dominant, with a compositional ratio of 9 pine to 1 spruce. During this 5-year drought, the number of spruce decreased from 1076 to 83 (from 1935 to 1941). In the following years, there was a further loss of spruce. By 1951 the plot had only isolated spruce trees (6 in total).

Extreme climatic factors altered the forest ecosystems' structure; the mixed pine and spruce composition has transformed into pure pine structure (10 pine trees). The loss of spruce led to a dense growth of second tier of hardwoods: birch, maple, oak and linden, which are present today and are currently affecting the soil-forming processes. The growth of these hardwoods influences soil reactions by providing more organic matter and cindery nutrients, making the soil more acidic. This enhances the humus-accumulation process and generally improves the soil quality for forests.

Alternatively, the long-term growth of mixed spruce and pine stands drastically slowed down the process of humus accumulation and reduced the concentration of major nutrients [3]. Figure 1 depicts the morphology of the soil profile in this plot based on the descriptions from 1909–1910 [6]. In 1984, a soil profile was studied here again by Naumov and Polyakov [5]. The morphological description of this study is presented in Fig. 2.

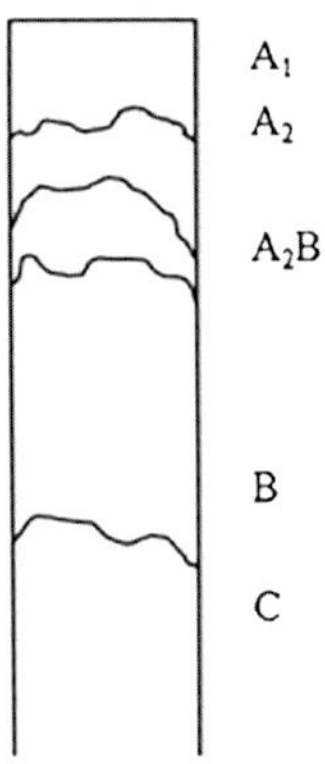

* Humus thickness of 14 cm
Net podzol thickness of 21 cm

Podzol with brown-ocherous spots with thickness of 15 cm

Red-brown loam with boulders, sand inclusion and clayveinsfrom 0.5 m to 1.4 m depth

Red loam of 1.5 m thickness with rare boulders and sand inclusions

Fig. 1. The morphology of the soil profile from the plot "Щ" (quadrant 4) based on the descriptions from 1909-1910. A letter designation of the horizons was given by the authors. * Due to the growth of pine here, the presence of litter was assumed

A_0	0-1 / 1	cm	The litter of fallen branches, needles
A_1	1-9 / 8	cm	The humus horizon is dark gray, light silty loam, loose, fresh with abundant roots, worms present; the transition to the next horizon is gradual
A_1^1	9-22 / 13	cm	The humus horizon is brownish-gray, light silty loam, fresh, with abundant roots, worms, lumpy, low compaction; the transition to the next horizon is gradual
A_2	22-39 / 17	cm	Podzolic horizon is light brownish-fawn with reddish spots, light sandy loam, slightly compacted, lumpy-platy structure with lots of roots, iron-manganese concretions, smears, pebbles; the transition to the next horizon is with streaks
A_2B	39-60 / 21	cm	The transitional horizon is of rusty-brown color with bluish veins, medium sandy loam, SiO_2 spots, lumpy-nutty structure, compacted, iron-manganese nodules, lots of roots; the transition to the next horizon is gradual
B 60…		cm	Illuvial horizon is rusty-brown, medium sandy loam, nutty structure, compacted, rare roots, iron-manganese concretions

Fig. 2. The morphology of the soil profile from the plot "Щ" (quadrant 4) in 1984. Type of soil is loamy deep soddy and shallow podzol soil on moraine clay loam

The morphological analysis for the one hundred-year period of the soil profile showed an increase in the thickness of the humus layer from 14 to 21 ± 1 cm and reduction in the depth of the podzolic layer from 21 to 17 ± 0.8 cm [2]. The illuvial horizon depth was found to have decreased from 50 to 60 ± 1 cm: a positive and very important change for the tree plantations. This illuvial horizon is mainly characterized by high bulk density, making it difficult for plant roots to penetrate and therefore, impacting the nature of the root system.

Comparison of the observed changes in the thicknesses of the humus and podzolic horizons that reflected the dense overgrowth of hardwoods after the drought of 1938–1939 with the changes observed in research period from 1941 to 1984 allowed us to estimate the rate of these changes. The thickness of the humus horizon increased at the rate of 0.16 cm/year and the podzolic horizon was reduced with the rate of 0.09 cm/year in pine stands due to the effect the succession of the hardwoods linden [5]. It is therefore determined, that these indicator scan be used to monitor the activity of sod and podzolic processes for this type of the tree plantations.

3 Ecological Characteristics of the Plot "Ъ", Quadrant 4

The plot "Ъ" of quadrant 4 was established in 1886 by Dr. M.K. Turski in the north-eastern part of the FES (next to the plot "Щ") to study the growth of naturally-occurring pine trees of natural origin [6]. Silvicultural characterization of the plot is presented in Table 2. The composition of the phytocenosis in this plot consisted of pure pine (P) stands (10 pine trees) until 1929 [5] By that time, the second succession maple (M), linden (L), and oak (O) grew in, resulting in the following composition $\frac{10P}{6M4L}$. Due to favorable conditions, by 1944, oak and lindens grew successfully [5] resulting in the composition $\frac{10P}{7L3O}$.

Table 2. Silvicultural characterization of the plot "Ъ" (quadrant 4) of the FES

Year of recount	Age, years	Tree species	The number of trees, units/ha	Average		Composition	Overall productivity, m^3/ha
				Diameter, cm	Height, m		
1896	26	Pine	4458	9.6		10 P units B	undefined
		Birch	183	9.1			
1904	34	Pine	3212	12.1	9	10:00 PM	— » —
1909	39	Pine	2295	14.3	12	10:00 PM	— » —
1916	46	Pine	1575	17.4	16.5	10:00 PM	— » —
1924	54	Pine	1343	18.8	17.5	10:00 PM	— » —
1929	59	Pine	1282	18.7	18.8	10P	— » —
		Maple	1819	4		6M4L	
		Oak	220	3.7			
		Linden	61	14.3			
		Birch	12	3.2			

(*continued*)

Table 2. (*continued*)

Year of recount	Age, years	Tree species	The number of trees, units/ha	Average		Composition	Overall productivity, m^3/ha
				Diameter, cm	Height, m		
1939	69	Pine	830	22.8	21.5	10:00 PM	570
		Maple	586	5.6		7L3O	
		Oak	98	6.2			
		Linden	49	18.4			
		Birch	12	7			
1944	74	Pine	684	28.3	22	10:00 PM	579
		Maple	12	9		7L3O	
		Oak	73	9.8			
		Linden	49	21.8			
		Birch	12	10			
1949	79	Pine	610	24.1	22.5	10:00 PM	594
		Oak	49	17.3		7 L3O	
		Linden	49	24.4			
		Birch	12	12			
		Maple	6				
1954	84	Pine	586	24.8	23	10:00 PM	526
		Oak	49	18		7 Ln 3O	
		Linden	49	26.8			
		Birch	12	16.3			
1959	89	Pine	420	25.7	23	10 P 7 L 3O	634.5
		Oak	49	19.3			
		Linden	49	28.7			
		Birch	12	19			

These observations lead to the determination that the emergence and activity of hardwoods (linden and maple) at the end of 1920s contributed to the weakening of the podzol-forming process especially by maple (which was in significant quantities; 1819 trees). This determination is supported by the analysis of the soil profile morphology conducted at the beginning and end of the 20th century (Fig. 3). The humus horizon increased from 19 cm in 1910 to 22 ± 0.7 cm in 1984.

The illuvial horizon depth diminished from 54 cm to 72 cm (1984). The admixture of deciduous species (linden, maple, and oak) led to an increase in root zone capacity and generally improved the structure of the soil profile and growing conditions for plantations. The depth of the humus layer during the growth of the linden, maple, and oak mixing in with the pure pine since 1929 allowed us to determine the rate of increase in the thickness of the humus layer due to this succession.

The year of 1929 was considered as the beginning of the change in the humus formation process, because prior to that year the pure pine stands did not have any effect on thickness of that horizon [5]. The difference in the thickness of the humus layer between 1984 and 1909 (equal to 3 cm), was used to obtain the rate of increase of the horizon of accumulation between those years, which was equal to 0.05 cm/year.

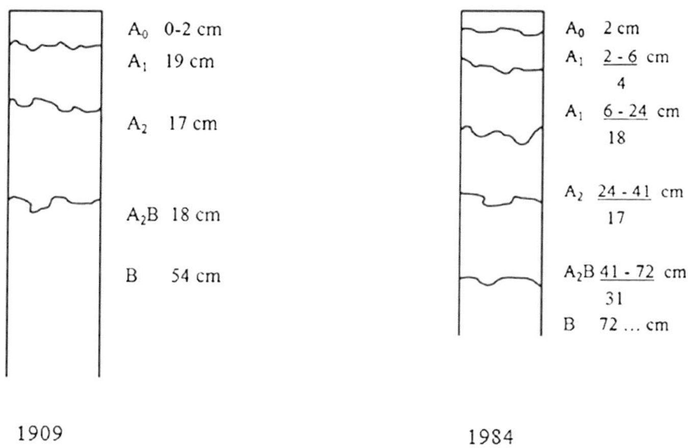

Fig. 3. Morphology of the soil profiles in the plot "Ъ" (quadrant 4) in 1909 and 1984

4 Ecological Characteristics of the Plot "Л", Quadrant 7

The plot "Л" of quadrant 7 was established in 1910 by Dr. N.S. Nesterov in the 80-year-old naturally-occurring pine stand with an admixture of birch (B) and a second tier of linden and 45–60-year old oak [5]. The composition of the forest stand was $\frac{8P2B}{7L3O}$. Silvicultural characterization of the plot is presented in Table 3. Figure 4 shows changes in the soil profile morphology in the plot "Л" (quadrant 7) from 1909 to 1984.

For the 80 year-period, the humus horizon increased from 15–16 to 23 cm ± 1.3 cm. The podzolic horizon was influenced by the active growth of hardwood (linden, oak) (markedly in 19380 and consequently transformed into a transition layer with a reduction in its thickness from 26 cm to 14 cm ± 0.7 linden [3]. During that period, the linden and oak matured from the second tier into the first one increasing productivity. In 1938, the wood stock of the stand per 1 ha was 86 m^3 and 18.9 m^3, respectively. In 1960, the stocks increased up to 143.5 m^3 and 27 m^3.

As the productivity of hardwood increased, functioning of coniferous (pine) phytocenoses weakened. According to Mosina et al. [3], the pine was in active growth for one hundred years (until 1938); after which it declined over the 20-year period from 1938 to 1960. This decline was marked by a decrease in stand density from 339.9 m^3/ha to 140 m^3/ha. The year 1938 was considered to be the beginning of the most pronounced period of the hardwoods weakening of the podzolic forming process, leading to an increase in the thickness of the horizon of accumulation. The rate of accumulation of the humus was determined from the layer's thicknesses in 1984 and 1938 (23 cm and 16 cm, respectively), and estimated to be 0.15 cm/year (7 cm per 46 years)

In adjacent plots of similar species, the rate of humus accumulation in the soil differed by a factor of 3 (0.05 cm/year and 0.16 cm/year). In the plots located approximately 150 m apart (quadrant 4, plot "Щ" and quadrant 7, plot "Л"), the rates of the humus accumulation were identical (0.16 cm/year and 0.15 cm/year,

Table 3. Silvicultural characterization of the plot "Л" (quadrant 7) of the FES

Year of recount	Age, years	Tree species	The number of trees, units/ha	Average		Composition	Overall productivity, m^3/ha	Stock, $м^3$
				Diameter, cm	Height, m			
1910	80	Pine	331	35.3 2	25	8P2B	undefined	Unde-fined
		Birch	156	4.6	22	7L3O		
		Linden	349	11.2				
		Oak	516	6.1				
		Spruce	4	12				
1913	83	Pine	331	36.3	25	8P2B	— » —	— » —
		Birch	135	26.2	22.5	7L3O		
		Linden	396	11.4				
		Oak	407	6.8				
		Spruce	4	13				
1931	101	Pine	233	38	25	8P2B	— » —	— » —
		Birch	109	29.6	23	8L2O		
		Linden	233	19				
		Oak	240	10.7				
		Spruce	7	11.4				
1938	108	Pine	236	39.4	25.5	8P2B	661.9	339
		Birch	98	30.7		8L2O		77.2
		Linden	247	22.3				86
		Oak	189	13.5				18.3
		Spruce	4	16				0.6
1945	115	Pine	185	38.3	25.5	9P1B	662.1	259
		Birch	44	30.4		8L2O		35.6
		Linden	236	22.2				81.8
		Oak	116	14.4				15
1950	120	Pine	164	39.4	25.5	9P1B 8L2O	714.3	235.7
		Birch	29	28.5				19.3
		Linden	222	26.1				110.3
		Oak	94	17.8				20.8
1955	125	Pine	142	36.7	25.5	9P1B	740.6	185.1
		Birch	25	30.3		8L2O		19.4
		Linden	204	28.6				124.5
		Oak	91	19.8				23.3
1960	130	Pine	109	38.6	25.5	9P1B	767	140.2
		Birch	18	32.1		8L2O		15.7
		Linden	204	30.4				143.5
		Oak	84	21.8				27.1

respectively). The proximal plots (quadrant 4, plots "Щ" and "Ъ"), had rates of humus-accumulative horizon differed by approximately three times (0.16 and 0.05 cm). These differences may be explained by the origin of the tree stand.

It was observed that the slowest rate of accumulation of the top horizon (the first 0.05 cm) occurs in the forest of naturally-occurring pine. In the old-growth coniferous stands, whose composition was unchanged, the humus-accumulative process was slow.

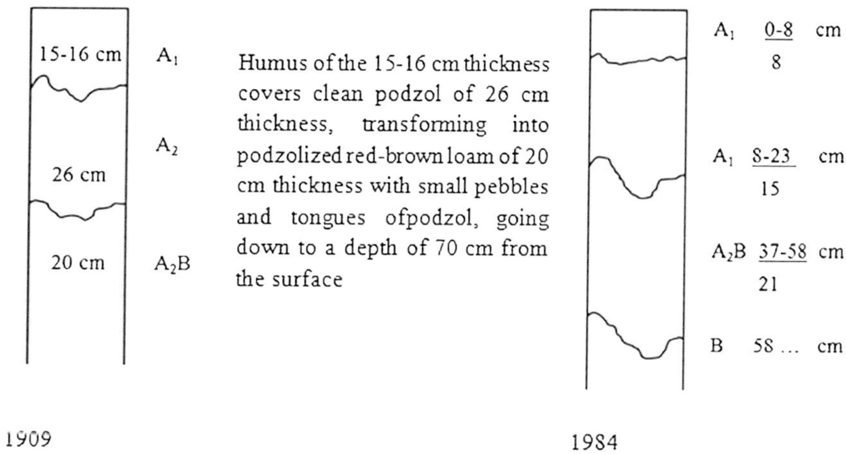

Fig. 4. Changes in the soil profile morphology in the plot "JI" (quadrant 7) from 1909 to 1984. The soil is light loamy deep turf deep podzolic on morainic loam. A letter designation of the horizons was given by the authors.

The authors acknowledge some arbitrariness of this method for evaluating the rate of humus accumulation. However, by using an identical method of comparison in other test areas, this allows comparison of the rates of the humus accumulation influenced by coniferous species in different growth periods.

The main contributor to the decline in the podzol-forming process and increase in the humus-accumulation process was the change in tree stand composition from conifers (spruce and pine) to hardwoods (birch, maple, linden and oak). Consequently, the composition the litter changed. The acidic decomposition products of coniferous species' litter were replaced with the products from hardwoods which have a larger base saturation.

Oak played a significant role in the decline of the podzol-forming process in these phytocenoses. The oak trees generated greater biomass of organic matter than the coniferous-deciduous species, and as a result, increased the amount of nitrogen and cindery elements [2]. Depending on the stand composition, the litter quantity and its cindery composition varied considerably along with other biochemical changes, which affected functioning of the soil-plant system. Oak played an important role in reducing the intensity of the podzol-forming process because of its ability to consume more water for transpiration compared to that of coniferous or mixed forests. This, in turn, also depressed the development of the eluvial process [3].

The deciduous trees have different growth requirements than conifers in terms of major nutrients and the effects on the soil; strengthening of the biological accumulation of mineral elements, nitrogen and humus accumulation. These differences account for diminished podzol-forming process. Coniferous trees take much less nutrients from soils than do deciduous trees [9]. This is especially obvious when comparing the consumption of the nutrients by the trees of the same age. For example, one oak tree at the age of 90–100 years takes up 169 g of calcium annually, whereas a pine tree of the same age takes 17 g, and a spruce - 21 g [8]. One oak tree consumes 102 g of nitrogen, whereas pine consumes 23 g, and spruce, 23 g.

The consumption of the nutrients from the soil (kg/ha) can be represented by the following increasing order:

$$\begin{array}{cccccc} \text{pine forest} & < \text{spruce} & < \text{oakry} & < \text{linden} & < \text{aspen} & < \text{birch} \\ & 62 & 73 & 133 & 162 & 258517 \end{array}$$

If the number of elements returned to the soil by litter and dead wood is expressed in terms of percent of consumption of these elements and compared among tree of the same age, then the following pattern can be identified. At the initial period of development (i.e. the sprucest 30–40 years), the return was 40–60% of the consumption of the elements. For the maturing and mature stands, a much smaller percentage of the elements was removed from the soil, and the return was 80–90% of the consumption. An exception was the oak forest, which returned to the soil about 70–80% of consumption during its first age class (1–20 years).

It can clearly be inferred that changes in the soil-forming processes, were associated with the changes of the stands from coniferous trees to deciduous. The change in vegetation led to an alteration in the ecosystem which was characterized by a variety of interrelated factors, such as the biochemical composition of the vegetation.

The Moscow region and Moscow City has a flush-type of the water regime where the research facilitates are located with corresponding biochemical composition of litter from different tree species affecting the intensity of the podzolic process as noted in this research. Different composition of the stand on the test plots and the observed the changes in soil properties, together, allowed, for the first time, to identify the soil evolution and its pace at the FES for a long-term period of 100 years. The composition of the main stands affected by the hardwood (linden, oak, maple, birch) is as follows:

$$6F\,4 \rightarrow 10P \rightarrow \frac{10\,\mathrm{P}}{\mathrm{BMLO}} \; \mathit{and} \; 10\,\mathrm{PB}(\mathit{few}) \rightarrow \frac{10P}{7\mathrm{L}3O} \tag{1}$$

The rate of the humus accumulation in these hardwood stands approximately 0.15 to 0.16 cm/year. The rate of decrease of the podzolic horizon was 0.09 cm/year. In similar stands, where coniferous (pine) trees were old-growth (over 100 years old) and of natural origin, the rate of formation of the humus horizon was measured to be 3 times lower than hardwoods; at the rate of 0.05 cm/year.

Increased formation of the humus in the soil can contribute to an evolution of the tree stand compositions; changing from stands of conifers, that are highly sensitive to environmental pollution, to more resistant deciduous species. This compositional change will have a significant positive effect on the environment of the northern district of Moscow and the entire city. The adaptation will allow the phytocenoses to provide highly productive ecosystem services due to the trees better suiting their environment and improving the soil quality.

Even with a longer growing season of evergreen coniferous stands, the intensity and functionality of coniferous phytocenoses is much lower compared with the deciduous stands. In the studied region, the length of a growing period for hardwoods is about 6 months; from mid-April to mid-October. The coniferous trees have a very weak resistance to the industrial and car emissions in comparison with the deciduous species [3]. The results of the study allow the prediction of the soil-forming processes, the state of the forests, and therefore, the surrounding environment, which is extremely important for soil-ecological monitoring in recreational forests and for ecological landscape mapping.

5 Conclusions

The composition of tree species and their substitution has an effect on the main soil forming processes in the profile of sod-podzolic soils, i.e. eluvial process and process of humus accumulation. As a result, it influences the thickness of the main genetic horizons of the soil profile.

Weakening of podzolic forming process was revealed as a result of the change in stands from coniferous to deciduous species (birch, maple, linden and oak).

The rate of formation of the humus horizon and the reduction of the thickness of the podzolic horizon was calculated. The rate change was due to the compositional change of the pine and spruce stands: 6 S 4 P (0.09 cm/year) within the mixed two-tiered plantations. Birch, maple, linden and oak trees effected this change.

Growth of native pine tree enhanced the podzol process resulting in an admixture of deciduous trees (linden, maple and oak) which reduced the rate of accumulation of the upper horizon by a factor of 3; from 0.16 cm/year to 0.05 cm/year.

The thickness of the humus horizon under the mixed conifer-deciduous stands was found to be thickened under the hardwoods.

Comparison of the taxation parameters with the soil characteristic will help in developing criteria for the crop stability and productivity in urban and suburban areas.

Acknowledgements. The publication was prepared with the support of the "RUDN University program 5-100".

References

1. Burkhard, B., Kroll, F., Muller, F., Windhorst, W.: Landscape's capacities to provide ecosystem services - a concept for land-cover based assessments. Landsc. Online **15**(2009), 1–22 (2009)
2. Mosina, L.V., Dovletyarova, E.A.: Microbiological diagnostics of problematic environmental situations at the recreational natural facilities. In: Bulletin of Peoples' Friendship University of Russia. Series «Agronomy and animal industries», №. 5, pp. 130–140. PFUR (2013)
3. Mosina, L.V., Dovletyarova, E.A., Andriyenko, T.N.: The forest experimental station of Russian State Agrarian University - Moscow Timiryazev Agricultural Academy as object of environmental monitoring of forest and forest-park landscapes of the megalopolis of Moscow, p. 221. PFUR (2014)

4. Mosina, L.V., Dovletyarova, E.A., Petrovskaya, P.A.: Microbiological assessment of a condition of forest and forest-park ecosystems. In: Bulletin of Peoples' Friendship University of Russia. Series «Agronomy and animal industries», №. 4, pp. 42–51. PFUR (2015)
5. Naumov, V.D., Polyakov, A.N.: 145 years of the forest experimental station of the Russian State Agrarian University of Moscow Agricultural Academy (RSAU-MAAT), p. 511 (2009)
6. Nesterov, N.S.: Forest Experimental *station* in Petrovsky – Razumovsky. In: OGIZ, Sel'khozgiz, p. 35 (1935)
7. Percy, K.E., Ferretti, M.: Air pollution and forest health: toward new monitoring concepts. Environ. Pollut. **130**(1), 113–126 (2004). Original Research Article
8. Tkachenko, M.E.: The impact of certain forest species on soil. In: Pochvovedenie. №. 10 (1939)
9. Vitousek, P.: Nutrient cycling and nutrient use efficiency. Am. Nat. **119**(4), 553–572 (1982)
10. Wojciech, D., Bytnerowicz, A.: Monitoring environmental pollution in Poland by chemical analysis of Scots pine (Pinussylvestris L) needles. Original Research Article. Environ. Pollut. **87**(1), 87–104 (1995)

Use of the Field Data for Assessment of Hazardous Concentration of Pollutants in Soil and Modelling of Species Sensitivity Distribution

V.K. Shitikov[1(✉)], A.E. Ivanova[2], K.A. Kydralieva[3], and V.A. Terekhova[2,4]

[1] Institute of Ecology of the Volga Basin, Russian Academy of Sciences, 10 Komzina, Togliatti 445003, Russia
stok1946@gmail.com

[2] Lomonosov Moscow State University, 12-1 Leninskie Gory, Moscow 119991, Russia
vterekhova@gmail.com

[3] Moscow Aviation Institute, Volokolamskoeshosse, 4, Moscow 125993, Russia

[4] Severtsov Institute of Ecology and Evolution, Russian Academy of Sciences, 33 Leninskij pr., Moscow 119071, Russia

Abstract. A rapid assessment of Environmental Quality Criteria and probability of Ecological Risk without special toxicometric experimenting is an actual problem in environmental science. The article presents a statistical prediction methodology of approximated no-effect concentrations (NOEC) and modelling of Species Sensitivity Distribution (SSD) based on the field observation data. We use values of species abundance of tested community, which were located on a set of sites of region under study with wide variation range of polluting substances concentration. Statistical processing includes the following sequence stages: (1) calculation of distances matrix in multidimensional species' space between each pair sites; (2) nonmetric multidimensional scaling (NMDS) is applied to reduce a 2-dimensional plot matrix of sites and species projections; (3) the analyzed contamination mediafactors are interpreted as an ecological gradient in species compositions and construction of the additional ordination axes; (4) generalized additive models (GAM) are build and 3D smoothing surfaces of spatial distribution of pollutant's concentration on ordination plot are fitted; (5) using the fitted models predicted values of PV ecological maxima and the upper boundary values of TV confidence intervals of each species for each single compounds are found; (6) obtained data are used for SSDs modelling. The methodology has been supported by results of bioindication for communities of microscopic fungi of soil samples from the former uranium mining province (Kyrgyzstan). Threshold values of six soil contamination indicators that ensure a pre-given admissible probability of environmental risk have been determined.

Keywords: Bioindication · Technogenic soil contamination · Fungi communication · Nonmetric multidimensional scaling (NMS) · Species sensitivity distribution (SSD) · Risk assessment

V.I. Vasenev et al. (eds.), *Megacities 2050: Environmental Consequences of Urbanization*, Springer Geography, https://doi.org/10.1007/978-3-319-70557-6_15

1 Introduction

Contemporary methods of environmental risk assessment are based on the detection of critical impact levels, which are interpreted as a starting point of decrease in ecosystem stability, therefore disappearance of the basic structural components of *biocenoze* or destruction of functional communications begins [1, 2]. The basic way of hazardous concentration estimation of technogenic pollution with reference to real populations in environment where active experiment cannot be realized, is practical use of available laboratory-derived toxicity values for surrogates of species. The problems of correct extrapolation of toxic effect are distinctions in taxonomic structure, specific life-cycle stages, levels of the biological organizations, set of accompanying physical and chemical parameters of media, type of exposure temporal regimen, spatial characteristics [3].

Analysis of Species Sensitivity Distribution (SSD) is one of the statistical extrapolation methods of laboratory data on different natural environments [4–8]. The SSD curve approximates from a panel of available acute or chronic toxicity values (as a rule, LC_{50}), or other effect measures for different species with respect to a particular chemical and interpreted as an integral function of some theoretical distribution of probability. It was originally developed for the risk assessment of single substances through the setting of thresholds: a hazardous concentration affecting p % of species (HC_p, i.e. either p-th % quantile of received distribution), or the fraction of species potentially affected by a certain concentration [5]. For example, if the threshold concentration is assumed as HC_5, it means that it is hazardous (lethal) for 5% of the most sensitive species and neutral for the rest, i.e. the zero-hypothesis about absence of harmful exposure is accepted at 95%-s' level. Probabilistic environmental risk assessment can be generally presented as a distribution ratio of the exposure (observed) and sensitivity (predicted on SSDs) concentration of pollutants [9, 10].

From the beginning of the use of SSDs the importance and difficulty of laboratory-to-field extrapolation and possible bias of risk assessment, caused by these reasons, has been discussed [3, 11]. The most important differences include a whole range of phenomena: bioavailability, spatial and temporal variance in field exposures, genetic or phenotypic adaptation and etc. SSDs, as a matter of fact, in any way does not use the information on ecology of communities (interspecific interactions, trophic communications, habitat factors, or the specific importance of keystone species and functional groups).

Another source of bias caused by data selection is that the species used for toxicity testing are not a random sample from the community of species to be protected [12] or in general with it not to coincide [13]. Often there is only the very narrow range of species tested relative to communities potentially exposed [14]. For example, micromycete and microbial communities in soil are almost inevitably under represented of toxic values when SSDs are intended to include them. Massive laboratory-derived determination of toxicometric indicators (NOEC, EC_{50}) for diverse ecotoxicants and with respect to multiple soil biota types is not actually feasible.

Therefore actual problem is search of the approaches that would enable a rapid assessment of the soil environmental condition basing merely on field observation data without special toxicometric experimenting.

The factorial ecology considers ways of statistical fitting of function of distribution of an abundance of any specie on a gradient of change of studied factors of environment, such as the maintenance of chemical substances, availability of resources etc. [15–18]. The simplest models of such dependences known as Species Response Curves, estimate three major parametres: optimum position, its amplitude and width of the response – see Fig. 1. The optimum defines preferable value PV of the factor where the specie can be found with the greatest probability, that is, localization of peak of distribution. Tolerance is connected with ability of population of a species to live and reproduce posterity in not optimum environment. The tolerant interval estimates a factor range in which the basic indicators of physiological activity or abundance of population can be remained or restored. Its TV right boundary value actually corresponds to maximum NOEC inefficient concentration, and any exposure exceeding this threshold, are considered as the hazardous. However basic difference of these estimations of ecological parametres from toxicity values LD_5 consists that they consider all set of conditions of a concrete habitat.

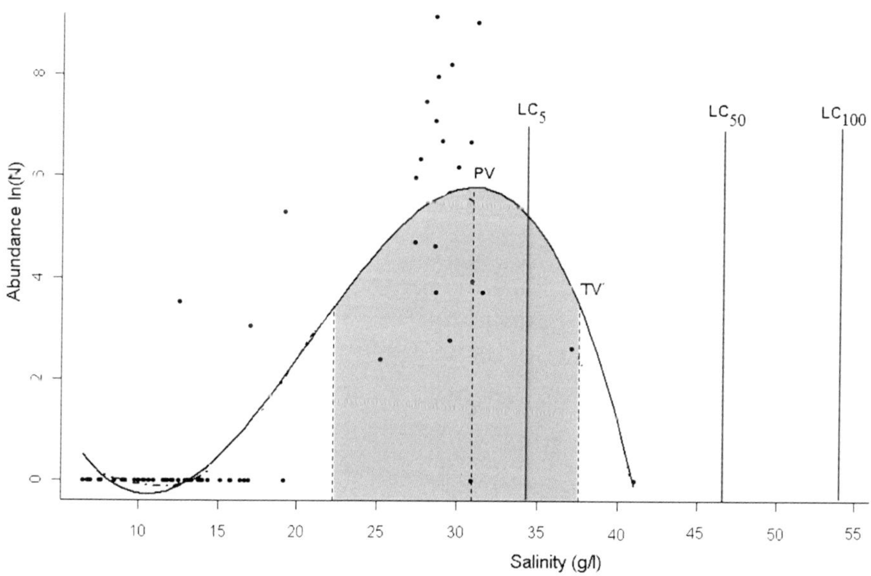

Fig. 1. Curve of distribution of abundance *Palpomyia* sp. on a gradient of salinity of water in inflows of the lake Elton. PV – preferable value of salinity; the grey fills in a range of tolerance with right boundary value TV; toxicological indicators LC_5, LC_{50}, LC_{100} are presented roughly.

In this article we propose an SSD alternative which takes into consideration absence of toxicological values by species of studied community, and uses in the analysis only raw data of field researches from a limited number of observation points. We consider possibility to utilize ordination procedures and multidimensional smoothing models for the estimation of a preferable PV and tolerance TV values for each species. Additionally, we propose a probabilistic risk of decrease of taxonomic richness estimation that links the modeled species distribution with the variability of environmental exposure conditions. The applicability of the proposed methods is elucidated in a case study on response for communities of microscopic fungi based on assessment of environmental risk of soil contamination from past uraniferous ore production.

2 Materials and Methods

2.1 Data Set for Illustration

Comprehensive studies of soil contamination were performed in the area of Kadzhi-Sai settlement (Kyrgyzstan), where low uranium concentration ore deposits were developed back in 1947–1965. Sampling from top soil layers took place in May, 2014 from sites located both in the territory of uranium mine tailings, and in rather clean areas on the slopes of adjoining hills, on the Issyk Kul lake shore, and in Boomsky canyon (42°08' 48" N, 77°11' 10" E).

A level of technogenic impact on the area in terms of two groups of indicators was analyzed: the activity of three radionuclides (U-238, Ra-226, Pb-210) using spectrometer Canberra (USA) consisting of a germanium detector HPGe and 16 heavy metals and other chemical elements content in the top soil horizons using hand-held XRF spectrometer DELTA Classic (USA). Total soil contamination with heavy metals (Zc) was derived using the modified Pollution Load Index (PLI) as the geometric mean [19]:

$$Z_c = n(K_1 \cdot K_2 \cdot \ldots \cdot K_n)^{1/n} - (n - 1), \tag{1}$$

where n is the number of the components, $Ki = Ci/ Cib$, Cib and Ci is the background and actual content of the i-th element in the soil. To account for different toxicity of the elements local PLI-indexes for three classes of hazard were calculated separately: for high hazard $Z_{c(1)}$ (As, Cr, Pb, Zn), for moderate $Z_{c(2)}$ (Co, Mo, Cu) and low hazard $Z_{c(3)}$ comprised of background and rare earth elements (Ba, Ti, Fe, Mn, Sr, K, Ca, Rb, Zr). The summary PLI-index was calculated with allowance for correction factors for toxicity:

$$Z_c = 1.5\, Z_{c(1)} + 1.0\, Z_{c(2)} + 0.5\, Z_{c(3)} \tag{2}$$

Soil fungi are among the most extensive and diverse groups of organisms used for the biodiagnostics of an environmental condition of biotopes, for setting environmental standards, and for environmental risk assessment [20]. The results of bioindication studies of micromycete communities in the soil sampled from 4 sampling sites with disturbed habitats and from 3 sites located in relatively clean zones (control) were used to evaluate ecosystem's response. Isolation of the cultivated microfungi was performed by a standard procedure of water soil suspension plating from 1:100 dilutions to the Czapek agar medium in a 3-fold replication. Frequency (%) of occurrence of each species was presented as its share in soil subsamples, in which a particular species was isolated. In total, 41 microfungal species were detected.

2.2 Statistical Analysis

Statistical processing to assess critical exposure levels and environmental risk with a preset certainty was conducted in two stages: (1) using ordination methods, the calculation of contamination factors corresponding to maximal abundance of each fungal species, and (2) approximation of data from the theoretical distribution curve for probability of species occurrence.

Procedure of multidimensional ordination of communities consists in optimum projecting of the studied habitats on a plane with latent axes S1 and S2 [16, 21]. An matrix of frequencies of occurrence with 41 microfungal species from 7 sampling sites used as input data. The matrix **D** of distances in multidimensional species' space between each pair of the soil samples by the Bray-Curtis formula was calculated [22].

The ordination of microfungal communities was built by the algorithm of non-metric multidimensional scaling (NMDS). Then a minimum of "stress" Δ is searched, which reflects degree of distortion of mutual distances between sites at a reduction from initial multidimensional space to a 2-dimensional plot [23, 24]. The major advantage of NMDS method is that it does not require a priori any assumptions about statistical distribution from the input data in contrast to such approaches as analysis of principal components [21]. Further, the weighted average coordinates s_1 and s_2 were estimated for individual microfungal species on the NMDS projection, which identified their position relative to sampling sites, and the ordination plot of the species was built.

Environment factors were used for interpretation of ecological gradients in species compositions along the constructed additional axes which have been added to axes of unconstrained ordination. The disposition of these vectors on ordination diagram was defined by model of multiple regression, in which each factor of environment was used as a response, and coordinates of sites s_1 and s_2 - as explanatory variable. Significance of models is tested by permutation procedure.

For any of the analyzed soil contamination factors Y generalized additive model (GAM) was built and fitted 3D smoothing surfaces in the same ordination plot was added. Models looked like:

$$Y = \alpha + f_1(s_1) + f_2(s_2) + f_3(s_1, s_2) + \varepsilon, \tag{3}$$

where f_1, f_2, f_3– specially picked functions from the NMDS coordinates s_1 and s_2 in the form of smoothing polynoms or penalised *splines* with k freedom degrees [25]. Predicted values of ecological maxima $\mathrm{PV}_j = \hat{Y}_j$ corresponding to coordinates of the most probable position of each j-th species of fungi, j = 1, 2, …, 41, on the NMDS projection were found from the fitted models. So high Y values were taken as approximated tolerant threshold for the j-th species, that they were low probable within the limits of the smoothing GAM model, i.e. the upper boundary values of confidence intervals $\mathrm{TV}_j = \hat{Y}_j + t_{\alpha/2} S_{\hat{Y}_j}$, where $t_{\alpha/2}$ – quantiles of student's t-distribution at α = 95%, $S_{\hat{Y}_j}$ - standard prediction errors of regression.

Further on, the attained empiric distribution of the preferable species value PV_j and tolerance threshold value TV_j along the Y-axis of contamination indicator was approximated by the theoretical distribution of the continuous random variable. A choice of the best distribution from a set of possible candidates (normal, log-normal, Weibull, etc.) and estimation of its parameters were conducted basing on the likelihood function log maximum. Confidence intervals of the selected cumulative distribution function $F(\mathrm{PV}(t))$ and $F(\mathrm{TV}(t))$ were estimated by a parametrical bootstrap method [4, 26].

All calculations were performed using *vegan* package of statistical environment R v. 3.02 [27].

3 Results

3.1 Multivariate Analysis of Data

The observed variability of the species structure of micromycete communities is associated basically with the gradient of the environmental conditions change in the area under study. The ordination graph in Fig. 2a testifies to rather clear differentiation of the sampling sites: soil samples from the uranium mine tailings (2 and 3) and from the natural reserve in Boomsky canyon (14) occupied extreme positions on the main axis S_1 of the nonmetric projection. Variability of the microbiota structure in other habitats with an intermediate contamination level was determined by the second ordination axis S_2.

Provided coefficients of correlation between the soil contamination indicators and the coordinates s_1 and s_2 on the ordination axis are calculated, it is possible to plot additional axes of physical gradients reflecting the nature and power of each factor's impact. The arrows of factor loads shown in Fig. 1a are approximately close both in direction and in length, so soil contamination factors in the studied region are likely to form an interconnected and *multicollinearity* complex. The best correlation (R^2 = 0.83, p = 0.022) was noted between a variation of the fungal communities structure (by frequency of occurrence of the found out species) and the concentration of cobalt (Co, mg/kg) in soil, - see surface smoothing by the GAM model in Fig. 2a.

The ordination of microfungi species groups (Fig. 2b) is closely connected with the ordination of habitats. If a species is only encountered in one sample, its position on the graph coincides with a point corresponding to a sampling site. Otherwise the species position is determined by weighted average coordinates of its possible several habitats. We believe, that it is a point of an "ecological optimum", where species occurrence most probably.

If a 3D smoothing surface (3) is built for any of the analyzed soil contamination factors it is easy to calculate a preferable values PV_j and thresholds of tolerance TV_j for points of an optimum of each *j*th species, which can be used further for modelling of probabilistic distribution of sensitivity. For some species on Fig. 2b the calculated values are resulted in Table 1.

3.2 Statistical Distribution of Species Occurrence

Further calculations were performed for soil contamination indicators presented in Fig. 2. With the use of preferable values PV_j for 41 microfungal taxas, the parameters of the Species ecological Maxima Distribution (SMD) were estimated on the scale of

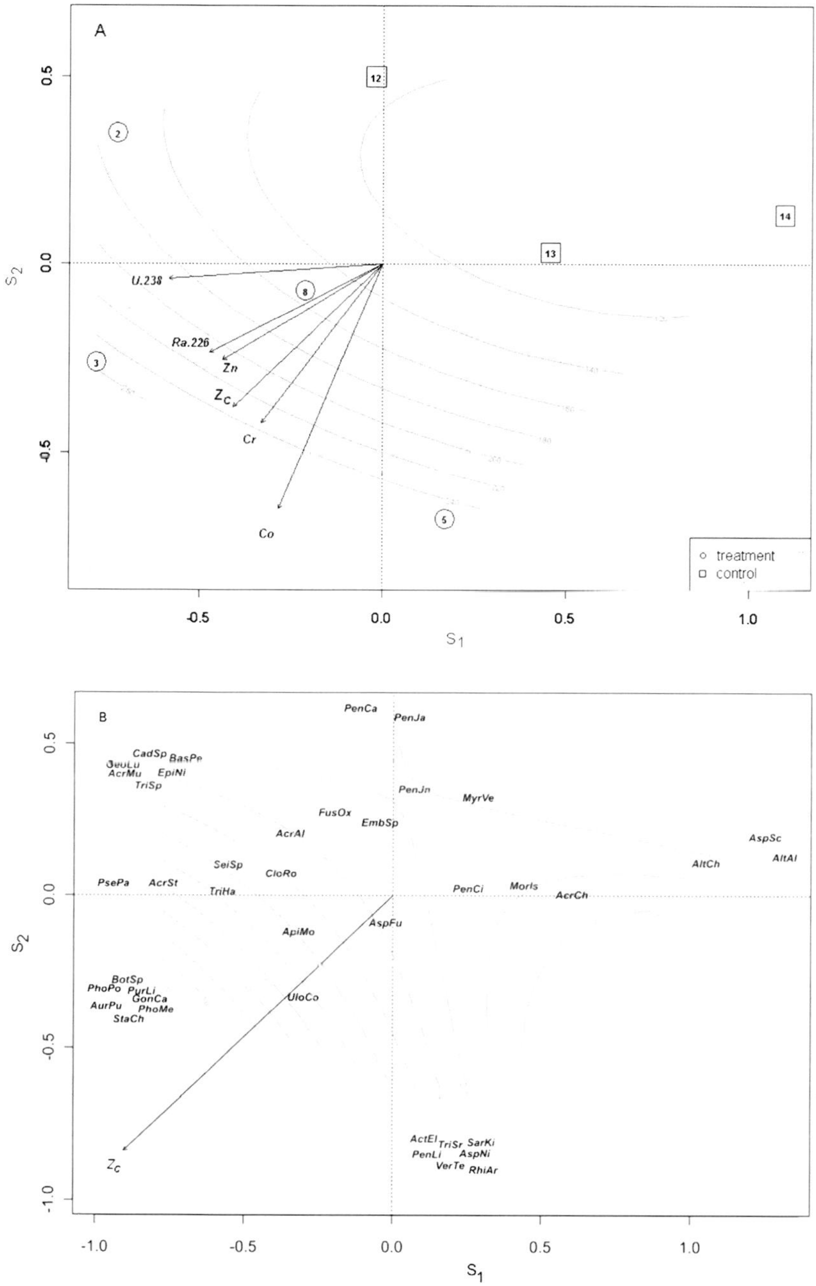

Fig. 2. Ordination of nonmetric multidimensional scaling data: (a) sampling sites (2, 3, 5 – a dump, 8 – a residential area of settlement Kadzhi-Saj, 12, 13 - the Issyk Kul lake shoreline, 14 – Boomsky canyon); (b) microfungal species (for some codes see Table 1). The arrows denote additional axes of physical gradients: index Zc, U-238 and Ra-226 radionuclide activity in soil, and the Co, Cr and Zn content. Grey isoclines show the cobalt content (2A) and Saet's index Zc (2B) calculated using the additive model.

Table 1. Coordinates s_1-s_2 on ordination plot (Fig. 2(b), preferable value PV of the concentration of cobalt (mg/kg soil), standard error of model and the right borders of a tolerant interval TV for some species of micromycete in soils of the former uranium-producing province

Codes	Species of fungi	s_1	s_2	PV (Co)	Error	TV (Co)
AcrCh	*Acremonium charticola* (Lindau) W. Gams	0.52	0.04	107.9	19.06	146.0
FusOx	*Fusarium oxysporum* Schlecht	−0.28	0.28	133.6	16.25	166.1
AltAl	*Alternaria alternata* (Fr.) Keissl	1.25	0.16	124.8	33.97	192.7
PenJa	*Penicillium janczewskii* K.M. Zalessky	−0.02	0.62	137.4	37.82	213.1
TriHa	*Trichoderma harzianum* Rifai	−0.58	−0.02	183.7	17.41	218.5
UloCo	*Ulocladium consortiale* (Thüm.) E.G. Simmons	−0.30	−0.36	215.9	18.87	253.6
AcrMu	*Acremonium murorum* (Corda) W. Gams	−0.82	0.43	183.2	35.71	254.7
PsePa	*Pseudogymnoascus pannorum* (Link) Minnis and D.L. Lindner	−0.85	0.04	211.3	23.67	258.7
StaCh	*Stachybotrys chartarum* (Ehrenb.) S. Hughes	−0.89	−0.33	287.9	34.29	356.5
VerTe	*Verticillium tenerum* Nees	0.20	−0.84	311.6	40.70	392.9

Table 2. Critical values of soil contamination indicators for various environmental risk levels (p %), calculated from the SMD and SSD curves (Fig. 3)

Indicator	Observed values (min ÷ max)	Preferable values PC_{15}	Hazardous values		
			HC_5	HC_{10}	HC_{15}
Cobalt Co, mg/kg	110 ÷ 261	130.3	141.3	160.0	176.5
Chromium Cr, mg/kg	15 ÷ 362	20.6	24.0	37.5	53.4
Zink Zn, mg/kg	11 ÷ 382	11.3	12.1	22.31	36.2
PLI-index Z_c	1.12 ÷ 20	1.94	3.5	5.2	7.1
U-238 activity, Bk/kg	24 ÷ 145	34.8	64.5	75.3	85.1
Ra-226 activity, Bk/kg	24 ÷ 134	34.4	81.2	91.9	101.4

each analyzed factor. Similarly with use of thresholds of tolerance TV_j the Species Sensitivity Distribution (SSD) were fitted. In all cases, the highest-likelihood approximations followed the *log-normal distribution law*.

Exemplary SMD and SSD curves are shown in Fig. 3 where it is possible to see how the response of microfungal communities is varying under the impact of different factors. The occurrence distribution over radionuclide activity and cobalt content scales is rather regular whereas the sensitivity in relation to other heavy metals and to the PLI-index Z_c (2) has a more contrastive nature. Of general pattern is considerable reduction of the specific richness and diversity of soil microfungal communities under the impact of heavy metals. However, in Fig. 3, it is easy to single out groups of the species possessing elevated resistance to some pollution forms unusual for normal conditions.

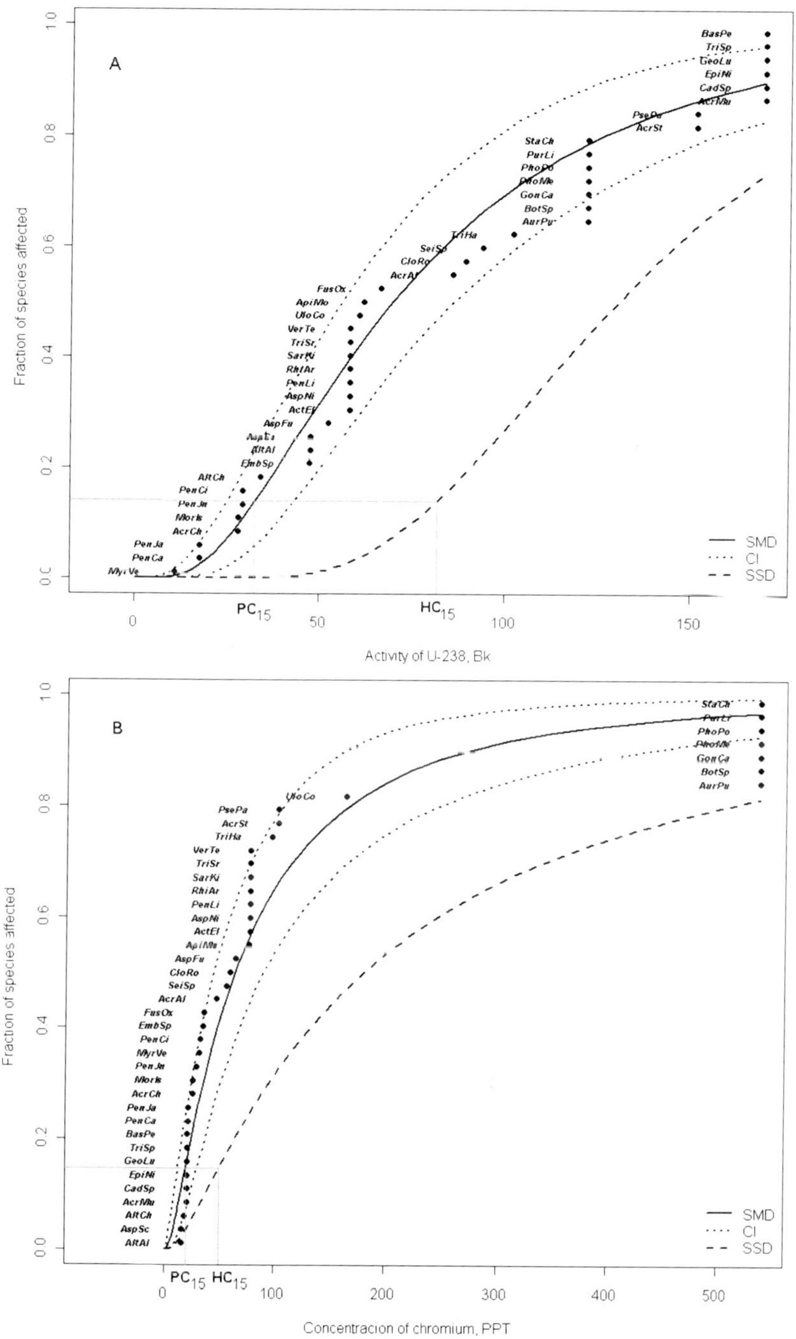

Fig. 3. Curves of log-normal distribution of probability of microfungal species maxima occurrence (SMD) and species sensitivity distribution at hazardous concentrations (SSD) on the scale of soil contamination indicators: 3A- U-238 radioactivity, Bk/kg, 3B– chromium content, mg/kg. CI – lower and *upper curves enveloping* the 95% confidence interval of SMD

If we take to arbitrary critical probabilities, e.g. $p = \{5, 10, 16, 20 \text{ and } 50\%\}$, using cumulative distribution curves SMD and SSD, it is possible to estimate a set of iso-effective values of, accordingly, preferable PC_p and hazardous for microfungal communities HC_p concentrations of the exposure factor (for examples see Fig. 3).

4 Discussion

Models of species sensitivity distributions (SSDs) were developed to derive criteria for the protection of biological entities in contaminated media. Assessment endpoints will vary depending on the protection goals and corresponds to a certain level of conservatism. Hence, it is necessary to define the relationship of the SSDs to sense of the setting of thresholds HC_p, given the input data. The acute LC_{50} values are based on mortality or equivalent effects (i.e., immobilization) on half of exposed organisms. At the ecosystems level at use SSDs it means, that at hazardous concentration HC_p in p % populations 50% of organisms will be lost approximately. The use of the SMD-curve determines a point of beginning of deviations from optimum of habitat conditions for p % of species and creates more stringent limitations to the estimation of critical concentrations. Assessment endpoints on the basis of thresholds of tolerance TV or no effective values NOEC, will occupy intermediate position (see Table 2) and make sense ``mild ecological hazardous'' for p % of populations as the probability of their resistance remains high.

To bring assessment endpoints into accord to the protection goals with reference to a concrete situation, the selection of values of the uncertainty factors (UF) and protection levels p is carried out [3–5]. Exist ambiguous opinions on what proportions p of the community or taxon as trigger values that should be considered as critically hazardous for an ecosystem [28]. Another uncertainty is the ambiguity of determining a share of maximum effect of impact p. This is usually done with account both of statistical "elasticity" of rated indicators and of a degree of researcher's responsibility for a conclusion (i.e. usually is a result of political compromise, instead of a science problem). Taking into account sense of thresholds of tolerance TV and ecological characteristics of tested community and media (micromicetes in soil), we believe that it is reasonable to be guided on p-values of 15%, which is in the precision region of ecotoxicologic methods [29].

The method of NOECs prediction and distribution of sensitivity of species on the basis of the field data and spatial models proposed by us is not the competitor to classical SSD. Use of available toxicity values is necessary and it is desirable, as the comparative analysis of results of modelling by various methods reduces uncertainty of assessmented endpoints. We will notice also a decrease in the specific richness is far from being a unique indicator for setting norms and standards in environmental regulations. Among the like is reduction of functional diversity or productivity, a switchover of the dominant species complex, etc. For decision-making it is important to have all accessible complex of the information on the response экосистемы in a gradient of influencing factors.

There are two reasons for regarding the cutoff values in Table 2 as tentative. First, ecological sense of NOEC in our case is identical to a finding of tolerant limits TV of existence of each species. Here again for a approximate estimation of the right tolerance boundary concerning an optimum we suggest to take advantage of confidential statisticans of smoothing model. Whether more correct estimation of tolerant ranges of occurrence of species is possible?

Theoretically, the TV estimation method has been only proposed for the normal or log-normal response (species abundance or occurrence) distribution on the exposure concentration scale (Oksanen et al., 2001). In more general case a finding of ecological optimum and tolerance ranges of occurrence is possible with the use of generalized regression models for each of species [30, 31]. However this would require several tens of measurements in identical ecosystem conditions. The wide range of a variation of concentration of polluting substance is necessary in any case.

Second, a limiting pollutant was not allocated, a combinatorics of cooperative impact from a mixture of toxicants was not considered, and also influence of accompanying parametres of environment, such as soil characteristics, was not analyzed. Problems connected with prediction of the potentially affected fraction of species and consequently for the risk assessment of chemical mixtures can be at least partially solved by various approaches [7]. Therefore we will notice only, that underestimated in comparison with environmental quality standards (for example, 100 mg/kg of the zinc) values of concentrations, presented to Table 2, we explain effect of additivity in a mix of components.

Let's pay attention to special circumstance, that the arrangement of points of sites on ordination plot in Fig. 2 is formal depends only on distances between them in multidimensional space of species. In turn, the configuration of points of species is defined by the sum of their statistical distributions under the influence of multiple stressors, uniting except influence analyzed pollutants, the combinations of other not considered factors of environment, including properties of soils [29, 32]. Hence, the prediction of preferable values PV_j for everyone individual stressor is carried out by the method described above against and taking into account influence of all of the others. Thus, rather than eliminating or minimizing extraneous variance, sources of variance may be explicitly acknowledged as part of the our SSD methodology.

Ultimately, one more problem of exposure rating is associated with high spatial heterogeneity of technogenic soil contamination, what, undoubtedly, tells on the number, abundance and distribution of fungal species [33, 34]. The redistribution of concentrations under precipitation impact within a microrelief, a relative height of sites, spotty nature of pollution, variability of the "*assimilation* capacity", biological activity of soil, etc. strongly influence bioindication results as a whole. Spatial interpolation of pointwise observations in geographical coordinates with a view to compensate for random fluctuations can be carried out by means of *kriging* models. We propose to perform smoothing surface modelling after the initial data projection onto a plane with latent axes directly related to the species structure of the community under study – that is to put aside natural spatial coordinates. In this case, the nonmetric multidimensional scaling method enables modelling of even and steady smoothing surfaces.

5 Conclusion

The modelling of Species Sensitivity Distribution allow to establish the critical (threshold) values of toxicant concentrations in the soils by using only field data without special toxicometric experimentation. In this work threshold values of six soil pollutants have been determined based on the analysis of the structure of micromycete communities of soil samples from the former uranium mining province (Kyrgyzstan).

Acknowledgement. This research was financed by the International Science and Technology Center (ISTC Project KR-2092) and Russian Science Foundation 14-50-00029 (isolation and identification of microfungi).

References

1. Brown, V.K.: Southwood TRE Secondary succession: patterns and strategies. In: Gray, A.J., Crawley, M.J., Edwards, P.J. (eds.) Colonization, Succession and Stability. Blackwell, Oxford, pp. 315–338 (1987)
2. Clements, W.H., Rohr, J.R.: Community responses to contaminants: using basic ecological principles to predict ecotoxicological effects. Environ. Toxicol. Chem. **28**, 1789–1800 (2009). doi:10.1897/09-140.1
3. Solomon, K.R., Brock, T.C.M., De Zwart, D., et al. (eds.): Extrapolation Practice for Ecotoxicological Effect Characterization of Chemicals. SETAC Press and CRC Press, Boca Raton (2008)
4. Newman, M.C., Ownby, D.R., Mezin, L.C.A., Powell, D.C., Christensen, T.R.L., Lerberg, S.B., Anderson, B.A.: Applying species-sensitivity distributions in ecological risk assessment: assumptions of distribution type and sufficient numbers of species. Environ. Toxicol. Chem. **19**, 508–515 (2000). doi:10.1002/etc.5620190233
5. Posthuma, L., Suter, I.I., Glenn, W., Traas, T.P.: Species Sensitivity Distributions in Ecotoxicology. CRC Press, Boca Raton (2001)
6. Gottschalk, F., Nowack, B.: A probabilistic method for species sensitivity distributions taking into account the inherent uncertainty and variability of effects to estimate environmental risk. Integr. Environ. Assess. Manage. **9**, 79–86 (2013). doi:10.1002/ieam.1334
7. Smetanová, S., Bláha, L., Liess, M., Schafer, R.B., Beketov, M.A.: Do predictions from species sensitivity distributions match with field data? Environ. Pollut. **189**, 126–133 (2013). doi:10.1016/j.envpol.2014.03.002
8. Del Signore, A., Hendriks, A.J., Lenders, H.J.R., Leuven, R.S.E.W., Breure, A.M.: Development and application of the SSD approach in scientific case studies for ecological risk assessment. Environ. Toxicol. Chem. **35**, 2149–2161 (2016). doi:10.1002/etc.3474
9. Verdonck, F.A.M., Aldenberg, T., Jaworska, J., Vanrolleghem, P.A.: Limitations of current risk characterization methods in probabilistic environmental risk assessment. Environ. Toxicol. Chem. **22**, 2209–2213 (2003). doi:10.1897/02-435
10. Scott-Fordsmand, J., Damgaar, C.: Uncertainty analysis of single-concentration exposure data for risk assessment - introducing the species effect distribution approach. Environ. Toxicol. Chem. **25**, 3078–3081 (2006). doi:10.1897/05-200R.1
11. Forbes, V.E., Calow, P.: Species sensitivity distributions revisited: a critical appraisal. Hum. Ecol. Risk Assess. **8**, 473–492 (2002). doi:10.1080/20028091057033
12. Suter II, G.W.: Comments on the interpretation of distributions in "Overview of recent developments in ecological risk assessment". Risk Anal. **18**, 3–4 (1998)

13. Smith, E.P., Cairns, J.J.: Extrapolation methods for setting ecological standards for water quality, statistical and ecological concerns. Ecotoxicology **2**, 203–219 (1993)
14. Cairns, J.J., Niederlehner, B.R.: Problems associated with selecting the most sensitive species for toxicity testing. Hydrobiology **153**, 87–94 (1987)
15. Austin, M.P., Cunningham, R.B., Flemming, P.M.: New approach to direct gradient analysis using environmental scalars and statistical curve-fitting procedures. Vegetatio **55**, 11–27 (1984). doi:10.1007/BF00039976
16. Jongman, R.H.G., ter Braak, C.J.F., van Tongeren, O.F.R.: Data Analysis in Community and Landscape Ecology. Pudoc, Wageningen (1987)
17. ter Braak, C.J.F., Looman, C.W.N.: Weighted averaging, logistic regression and the Gaussian response model. Vegetatio **65**, 3–11 (1986). doi:10.1007/BF00032121
18. Oksanen, J., Läärä, E., Tolonen, K., Warner, B.: Confidence intervals for the optimum in the Gaussian response function. Ecology **82**, 1191–1197 (2001)
19. Angulo, E.: The Tomlinson pollution load index applied to heavy metal "Mussel-Watah" data: a useful index to assess coastal pollution. Sci. Total Environ. **187**, 19–56 (1996)
20. Gadd, J.M.: Geomycology: biogeochemical transformations of rocks, minerals, metals and radionuclides by fungi, bioweathering and bioremediation. Mycol. Res. **111**(1), 3–49 (2007)
21. Legendre, P., Legendre, L.: Numerical Ecology. Elsevier Science B.V., Amsterdam (2012)
22. Wolda, H.: Similarity indices, sample size and diversity. Oecologia **50**, 296 (1981). doi:10.1007/BF00344966
23. Kruskal, J.B.: Multidimensional scaling by optimizing goodness-of-fit to a nonmetric hypothesis. Psychometrika **29**, 1–28 (1964). doi:10.1007/BF02289565
24. McCune, B., Grace, J.B.: Analysis of Ecological Communities. MjM Software Design, GlenedenBeach (2002)
25. Wood, S.N.: Generalized Additive Models: An Introduction with R. Chapman, Hall/CRC, Boca Raton (2006)
26. Davison, A.C., Hinkley, D.V.: Bootstrap methods and their application. Cambridge University Press, Cambridge (1997)
27. Oksanen, J., Blanchet, F.G., Kindt, R., et al.: Vegan: Community Ecology Package. R package version 2.0-2. (2011). https://cran.r-project.org/web/packages/vegan. Accessed 24 July 2016
28. Hopkin, S.P.: Ecological implications of "95% protection levels" for metals in soil. Oikos **66**, 137–141 (1993)
29. Efroymson, R.E., Will, M.E., Suter II, G.W., Hull, R.N.: Toxicological benchmarks for screening contaminants of potential concern for effects on terrestrial plants. ES/ER/TM-85/R3. Oak Ridge National Laboratory, Oak Ridge, TN, USA (1997)
30. Coudun, C., Gégout, J.-C.: The derivation of species response curves with Gaussian logistic regression is sensitive to sampling intensity and curve characteristics. Ecol. Model. **199**, 164–175 (2006). doi:10.1016/j.ecolmodel.2006.05.024
31. Jansen, F., Oksanen, J.: How to model species responses along ecological gradients – Huisman-Olff-Fresco models revisited. J. Veg. Sci. **24**, 1108–1117 (2013). doi:10.1111/vs1250
32. Comte, L., Lek, S., de Deckere, E., de Zwar, D., Gevrey, M.: Assessment of stream biological responses under multiple-stress conditions. Environ. Sci. Pollut. Res. **17**, 1469–1478 (2010). doi:10.1007/s11356-010-0333-z
33. Tropina, O.V., Terekhova, V.A., Semenova, T.A.: The variability of the structure of micromycete complexes as the result of soil heterogeneity. Mikologiya I Fitopatologiya **37**(6), 74–79 (2003)
34. Terekhova, V.A.: The importance of mycological studies for soil quality control. Eurasian Soil Sci. **40**(5), 643–648 (2007). https://doi.org/10.1134/S1064229307050158

Chemodiagnostic by Lipid Analysis of the Microbial Community Structure in Trace Metal Polluted Urban Soil

V.A. Terekhova[1,2(✉)], N.V. Verkhovtseva[1], M.A. Pukalchik[1], I.R. Vodolazov[1], and V.K. Shitikov[3]

[1] Lomonosov Moscow State University, 12-1 Leninskie Gory, Moscow 119991, Russia
vterekhova@gmail.com
[2] Severtsov Institute of Ecology and Evolution, Russian Academy of Sciences, 33 Leninskij pr., Moscow 119071, Russia
[3] Institute of Ecology of the Volga Basin, Russian Academy of Sciences, 10 Komzina, Togliatti 445003, Russia

Abstract. This study is focused on assessment of microbial community structure differences in urban soils by means of lipid analysis. Soil samples (0–20 cm deep topsoil layer) contaminated with trace metals to various extents were taken from six sites located in the rural part of the city of Kirov (Russia). The samples were measured for pH and total Pb, Ni, Cr, and Cd. To study the microbial community, microorganism markers (fatty acids) were extracted, which were then diagnosed by mass spectrometry. Nonmetric multidimensional scaling plots of soil communities showed relative dissimilarity in control and polluted soils. The total lipid content in control samples was found to be significantly higher than in polluted soil. The highest indicator value was assigned to *Actinobacteria* phylum, whose concentration decreased remarkably in polluted samples, and anaerobes *Butyrivibrio* sp. and *Bifidobacterium* sp. were regarded as an indicator for soils with relatively low pollution exposure. Microbial profiles were also indicative of selective enrichment with competent species (*Desulfovibrio* sp., *Bacteroides fragilis,* and *Chlamydia* sp.) in soil greatly contaminated with heavy metals. This study suggests that the method of lipid diagnostic can be highly indicative of soil microbial structure and thus it can be used as a quantitative measurement of urban soil biological quality.

Keywords: Soil quality · Lipid markers · Microbial biodiversity · Heavy metal · Urban soil · Ordination · Nonmetric multidimensional scaling

1 Introduction

Urbanization is a dominant demographic trend and an important component of the global land transformation [9, 19]. Soil is a crucial constituent of urban ecosystems that plays an essential role in the cycling of elements and in biogeochemical as well as microbiological transformations acting as support for plants [20]. Primary information on heavily contaminated urban areas nowadays comes from many cities [4, 10, 17],

V.I. Vasenev et al. (eds.), *Megacities 2050: Environmental Consequences of Urbanization*, Springer Geography, https://doi.org/10.1007/978-3-319-70557-6_16

however chemical pollution assays are unable to evaluate the integrated biological effect representing a 'cocktail' of environmental factors acting simultaneously on biological systems. An analysis of microbial populations in urban soil can help detect signs of potentially genotoxic compounds even when they are not found by the chemical observation and provides crucial information on ecosystem functioning.

A widely accepted method of lipid analysis uses biomarkers which point at viable microbial biomass and shows dominant microbial diversity and microbial community 'fingerprints' [6, 13, 22]. Nowadays, a majority of authors focus on the characterization of a microbial community under various stabilization treatments [1, 3, 24], or in a model pollution study [2, 7, 8]. In spite of a great number of investigations devoted to phospholipid fatty acid analysis, little is known about urban ecosystem impacts on soil biota. There is a great need for improving the understanding of soil organism exposure in urban soils where direct causality between measured compound concentrations and effects on ecosystems is much more sophisticated.

This study focuses on assessment of the microbial community structure in urban soils by the lipid analysis. Basic and nonmetric multidimensional scaling statistical procedures were used to estimate dissimilarity between soil samples taken from relatively different pollution conditions.

2 Materials and Methods

2.1 Sampling Area

Soil sampling took place in May 2010 at six sites located in the center of Kirov city (58.3729N, 49.3743E), Russia. Its area is exposed to a humid continental climate with cold snowy winters and warm wet summers. The mean temperature and annual precipitation are 3.1 °C and 676 mm, respectively. These soils were selected because they had been neither fertilized nor replaced by artificial soil at least for the past 30 years, and existing traffic-related pollution is the key anthropogenic impact. In particular, sites A, B, C, D, and F are located in a gradient from 5 to 100 m from a road to a park area, while site G, though also in the park area, is free from traffic-related impact.

At all sites (1 × 1 m), samples were collected with quantization from the 0–20 cm deep top soil layer and each sample was properly mixed. A portion of each sample was kept refrigerated (4 °C) at the field moisture to be used in determination of soil biota. The remaining soil was air-dried and stored over several weeks in cardboard boxes at room temperature.

2.2 Chemical Analyses

Chemical soil analyses were conducted in 3 replicates. Soil pH was determined in a 1:5 w/v suspension in distilled water using a pH meter, and conductivity was measured by a conductivity meter. The moisture content was evaluated from weight loss after drying at 105 °C in an oven for 24 h.

The total heavy metals (Pb, Ni, Cr, and Cd) content were determined by digestion using a mixture of HNO_3, HF and $HClO_4$ according to [21]. A 0.5 g of 1-mm sieved soil were extracted with 5 mL of HF, 5 mL HNO_3 and 3 mL $HClO_4$, and the mixtures was evaporated to near dryness. The second portion of HF (3 mL), HNO_3 (3 mL) and $HClO_4$ (1 mL) was added, and the mixture was again evaporated. The residue was dissolved in 2 mol/L HCl (5 mL) and diluted to 25 mL with deionized water in a volumetric flask. The analyses of the elements were performed by means of Inductively Coupled Plasma Mass Spectroscopy (ICP – MS) at Agilent 7500a, USA.

2.3 Analysis of Lipid Components

From soil samples of 200 mg, the lipid components were extracted with acidic methanolysis (CM) in 0.4 ml of 1M HCl in methanol for one hour at 80 °C. At this stage, fatty acids (FA) and aldehydes were released from the lipids. As a result, FA was prepared as methyl esters (MESFA) and aldehydes as dimethyl acetals (DMA). Further, these components were extracted using 400 μl of hexane. The resulting extract was dried at 80 °C and the dried residue was treated with 20 μl of N, O-bis (trimethylsilyl) - trifluoroacetamide for 15 min at 80 °C to give trimethylsilyl ethers of hydroxy acids, alcohols and sterols. Then, 1–2 μl of the resulting reaction mixture was diluted with hexane to 100 μl and injected for analysis into the injector of the spectrometer. Trideutero-tridecanoate was used as an internal standard for quantitative measurements [14, 15].

2.4 Nonmetric Multidimensional Scaling

A method of nonmetric multidimensional scaling (NMDS) was used for a statistical analysis of the interplay between species' densities and a degree of the sampling site remoteness from a pollution source. An abundance matrix (millions of cells per gram) with 47 microorganism species from 6 sampling sites was used as input data. Such method gives an optimal projection of observed objects onto the surface with the NMDS1–NMDS2 coordinate axes using an arbitrary matrix of distances D between them. Then a minimum of "stress value" Δ is searched, which reflects degree of distortion of mutual distances between sites at a reduction from initial multidimensional space to a 2-dimensional plot. The major advantage of NMDS method is that it does not require any presumptions about characteristics of statistical distribution a priori from the input data [18] in contrast to the approaches such as a factor analysis and a method of principal components.

The abundance matrix was pre-converted using the $\chi 2$-transformation, which provides the most reasonable balance of population sizes of species in microbial community between taxons with high density and increased attention to complex of rare species [16]. The matrix of distances D between the areas in the multidimensional species' space was calculated using the standardized Manhattan metrics (Bray-Curtis measure).

Some species can be regarded as statistically significant ecological markers of soil pollution of particular groups of sites. The indicator value ($IndVal_{kj}$) of each j species for each group of sites k, which is one of K ones (K = 2 sample areas in our case) was determined. Indicator index is defined as the product of measure of species specificity A_{kj} and measure of fidelity B_{kj} [5]:

$$A_{kj} = N_{kj}/N_{+k} \quad (1)$$

$$B_{kj} = S_{kj}/S_{k+}; \quad (2)$$

$$\text{IndVal}_{kj} = A_{kj}B_{kj} \quad (3)$$

where in the formula for A_{kj} N_{kj} is the mean abundance of species j across the sites pertaining to cluster k and N_{+k} is the sum of the mean abundances of species j within the various clusters. In the formula for B_{kj} S_{kj} is the number of sites in cluster k where species j is present and S_{k+} is the total number of sites in that cluster. IndVal_{kj} is equal to 1 only if examples of specimen j were found in only one k set. The indicator value of species j for a partition of sites is the largest value of IndVal_{kj} observed over all clusters k of that partition: $\text{IndVal}_j = \max[\text{IndVal}_{kj}]$.

Based on an assumption of regular nature of association between abundance of species and particular group of soils it is possible to check a null hypothesis about randomnicity of association. The significance of the indicator value of each species is assessed by a following randomization procedure: if calculated p-value was bigger than 0.1, given species was regarded as background.

Statistical calculations were executed using *vegan* and *indicspecies* packages of R programming language.

3 Results and Discussion

3.1 Soil Chemical Properties

Urban soils were characterised by a slightly alkaline reaction of a soil solution (pHwater ranged from 7.51 to 8.06). The control soil reaction was close to neutral (7.40) whereas pH varied from 4 to 5.5 in natural sod-podzolic soils. The organic matter content was particularly high: total organic carbon (TOC) increased from 3.50 to 4.69% as moving closer toward the highway, while the background TOC level was 3.42%. Thus, all the samples fell under the high level category for sod-podzolic soils according to the conventional agrochemical classification. The elevated content of organic carbon in the samples can be apparently explained by the periodic introduction of peaty soil and by the flow of organic pollutants in the form of hydrocarbons. Also, high level enrichment with mobile forms of phosphorus (268–345 mg P_2O_5/kg of soil) and potassium (167–250 mg K_2O/kg of soil) was detected in the samples.

Mean concentrations of some metals in the topsoil of six sampling sites are shown in Fig. 1.

Soil contamination with heavy metals is an important factor for the ecological condition of soil. In terms of remarkable changes in the environmental quality standards for cars and automotive fuel, assessment of soil contamination with lead becomes highly relevant, meanwhile a necessity of taking into account pollutants such as nickel, chromium and cadmium still persists.

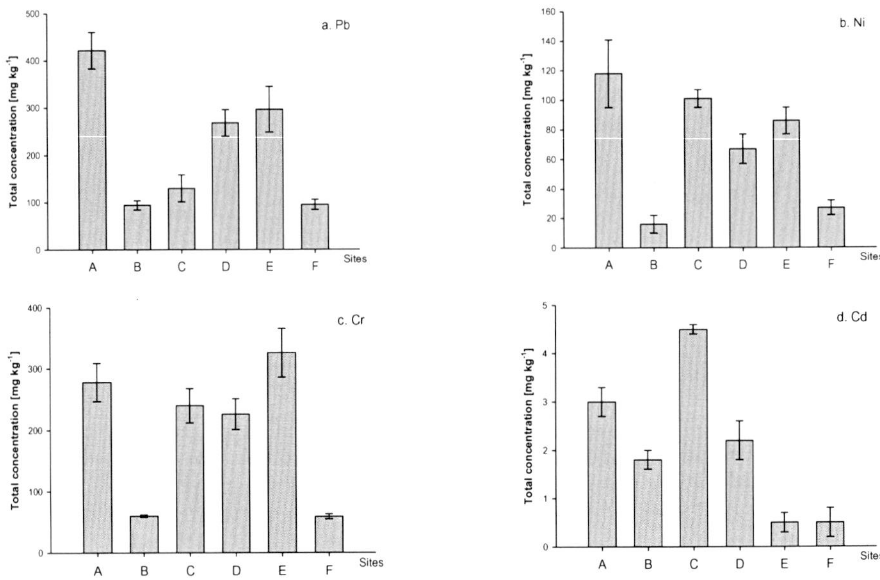

Fig. 1. Total metal concentrations in soils (Mean ± SD, n = 3).

3.2 Microbiological Analysis Using Lipid Markers

The microbiological analysis results for soil samples are summarized in Table 1. The values were calculated with a standard deviation of 15%.

Totally, five major bacterial phyla (with relative abundance no less than 1% in at least one sample) were found in urban soils, including *Proteobacteria, Actinobacteria, Bacteroidetes, Chlamydiae,* and *Firmicutis.* The diversity in the control sample (F) was

Table 1. The microbial community structure in soil particular sites (IndVal – indicator value of species, *p* – level of significance)

Taxonomic group (phylum)	Code of bacterium	Bacteria	The number (cell × 10^3 g^{-1}) per site						Cluster	Ind-Val	*p*
			F	A	B	C	D	E			
Proteobacteria	Acet	*Acetobacter* species	190	1760	2270	1700	2870	280	1	0.9	0.061
	AgRa	*Agrobacterium radiobacter*	0	40	390	170	280	0	1	1	0.082
	AeHy	*Aeromonas hydrophila*	0	510	40	420	260	110	1	0.84	0.14
	BrVe	*Brevundimonas vesicularis*	20	110	100	200	60	0	1	0.92	0.064
	Caul	*Caulobacter* species	0	370	1040	370	790	0	1	1	0.061
	Desu	*Desulfovibrio* species	20	80	330	480	470	0	1	0.97	0.064
	Meth	*Methylococcus* species	420	730	510	980	120	240	1	0.64	0.392
	Nitr	*Nitrobacter* species	0	270	670	320	530	40	1	0.96	0.066
	Ochr	*Ochrobactrum* species	0	0	0	0	230	0	1	0.25	1
	PsFl	*Pseudomonas fluorescens*	50	520	480	750	740	0	1	0.96	0.065
	PsPu	*Pseudomonas putida*	0	470	230	860	490	0	1	1	0.072

(continued)

Table 1. (*continued*)

Taxonomic group (phylum)	Code of bacterium	Bacteria	The number (cell × 10^3 g^{-1}) per site						Cluster	Ind-Val	*p*
			F	A	B	C	D	E			
	SpAd	*Sphingomonas adgesiva*	0	180	250	260	250	0	1	1	0.066
	SpCa	*Sphingomonas capsulata*	30	210	220	250	250	40	1	0.87	0.07
	Xant	*Xanthomonas* species	0	420	250	430	350	0	1	1	0.064
	Warb	WARB*	60	100	300	250	290	120	1	0.72	0.129
Actinobacteria	AcRo	*Actinomadura roseola*	0	100	240	310	290	0	1	1	0.072
	Bifi	*Bifidobacterium* species	150	280	880	0	0	240	2	0.4	0.885
	Cory	*Corynebacterium* species	0	340	0	120	140	0	1	0.75	0.318
	NoCa	*Nocardia carnea*	20	170	190	190	100	180	1	0.62	0.198
	Micr	*Micrococcus* sp.	0	610	830	1250	840	0	1	1	0.073
	Myco	*Mycobacterium* species	0	1400	275	4750	2990	0	1	1	0.066
	PrJe	*Propionibacterium jensenii*	0	2450	2260	2280	1910	0	1	1	0.07
	PrFr	*Propionibacterium freudenreichii*	0	1340	2830	0	3950	0	1	0.75	0.323
	Pseu	*Pseudonocardia* species	30	160	300	220	260	0	1	0.94	0.067
	RhEq	*Rhodococcus equi*	0	170	310	200	460	0	1	1	0.067
	RhTe	*Rhodococcus terrae*	720	800	380	710	520	0	1	0.63	0.476
	StNo	*Streptomyces/Nocardiopsis*	0	520	330	1080	1170	0	1	1	0.065
Bacteroidetes	BaFr	*Bacteroides fragilis*	0	0	0	0	30	0	1	0.25	1
	BaHy	*Bacteroides hypermegas*	0	40	30	40	40	0	1	1	0.067
	BaRu	*Bacteroides ruminicola*	0	240	150	240	110	0	1	1	0.082
	Riem	*Riemerella* species	0	190	210	330	190	0	1	1	0.062
	Cyto	*Cytophaga* species	0	270	120	250	120	0	1	1	0.08
	SpSp	*Sphingobacterium spiritovorum*	0	280	180	450	260	0	1	1	0.064
Chlamidiae	Chla	*Chlamydia* species	0	140	0	0	260	0	1	0.5	0.458
Firmicutes	AceF	*Acetobacterium* species	0	430	0	220	0	170	1	0.33	0.808
	BaSu	*Bacillus subtilis*	0	220	190	370	200	0	1	1	0.066
	Baci	*Bacillus* species	70	30	840	320	220	10	1	0.9	0.145
	Bu_2	*Butyrivibrio* species 1-2-13	200	10	390	210	150	270	2	0.55	0.727
	Bu_1	*Butyrivibrio* species 1-4-11	410	0	1560	90	310	580	2	0.5	0.737
	Bu7S	*Butyrivibrio* species 7S-14-3	2970	310	7800	1750	2020	4000	2	0.54	0.857
	EuLe	*Eubacterium lentum*	190	40	370	150	170	0	1	0.65	0.46
	Euba	*Eubacterium* species	0	180	780	1970	760	0	1	1	0.067
	ClPa	*Clostridium pasteurianum*	420	730	510	980	120	240	1	0.64	0.401
	ClPe	*Clostridium perfringens*	20	70	40	70	40	20	1	0.73	0.085
	ClPr	*Clostridium propionicum*	0	520	320	270	140	0	1	1	0.065
	Rumi	*Ruminococcus* species	110	1180	1250	2020	1370	130	1	0.92	0.069
	StMu	*Streptococcus mutans*	0	1460	1050	2250	2340	0	1	1	0.068

WARB* - *Wolinella-Acholeplasma-Roseomonas-Burkholderia*

found to be much lower than that of the other soils (A–D) (Fig. 2). The number of taxonomic groups in the most polluted sampled soils was significantly higher (50 species from 35 genera) as compared to control soils (23 species from 17 genera), which had been most probably caused by the known stimulation effect of new species development in the presence of small doses of pollutants. Anaerobic and facultative

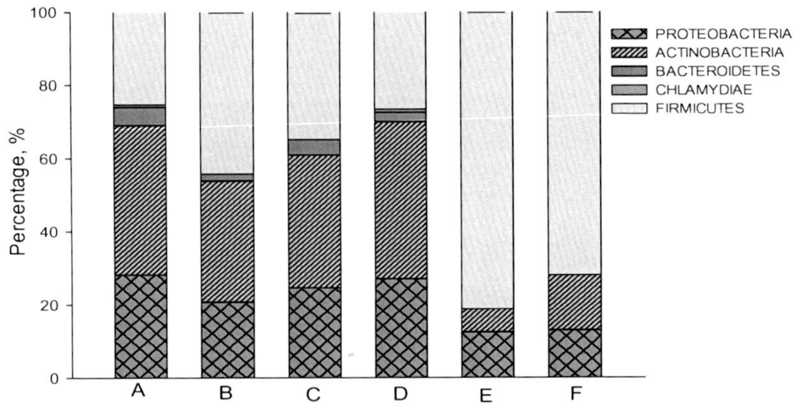

Fig. 2. Average relative abundances of the bacteria composition by phyla in each sample.

anaerobic species comprise a larger part of the microbial community in soils similar to the background. Their content in such soils was 20–25% higher than that within the polluted areas. Based on the enhanced accumulation of the marker taxonomic group *Mycobacterium* sp. in the samples characterized by a higher content of heavy metals, a conclusion can be made that the closer-to-the-highway areas were contaminated with hydrocarbons as well.

A proportion of actinomycetes from the genus *Streptomyces* which play a crucial role in the mineralization of organic substances, especially cellulose [23], increased along the gradient of the distance from the highway, apparently, due to an increase in diversity of plant residues. It went up from 27% (A) to 42% (D) whereas bacteria from the genus *Rhodococcus* decreased from 51% to 35% at the same places, which may be connected with biochemical features of these organisms such as a capability to transform and digest hydrocarbons. Moreover, *Streptomyces* representatives were absent in the background sample and in the fifth sample taken from the site most distant from the highway whereas *Rhodococcus* dominated in the community of the background sample (94%). This is indirect evidence of the fact that an organic substrate of the background sample (needles, twigs) is more available for primary processing by *Rhodococci* than by *Streptomyces*, which determined their exclusion from the microbial community. A higher content of *Actinobacteria* in the background community as compared to that of polluted soils, apparently, testifies to the sensitivity of this group to such kind of toxicants.

The applied method also allowed elucidation of potentially dangerous bacterial species in soils and analysis of their relation to the biotopic conditions. Thus, it was found that the contaminated areas contained 4–20 times more anaerobic specimens *Desulfovibrio* sp. known for hydrogen sulfide excretion. This toxic substance is able to suppress plant growth. A presence of the anaerobic specimen *Bacteroides fragilis* in the amount of 3×10^4 cells/g testifies to fecal contamination (D, Table 1). The same place was also contaminated with *Chlamydia* sp., whose pathogenic forms (*C. trachomatis*) are transmitted by birds and can cause severe eye and lung diseases. The content of chlamydia was one order higher than that of bacteroides -3×10^5 cells/g.

Seemingly, soils under the heavy metals load (A–D) were concurrently contaminated with hydrocarbons, which can be judged on the basis of elevated content of the indicator bacteria for such contamination, namely *Mycobacterium* sp.

The calculations of the biomass of micromycetes showed that it was 3.5–6.5 times lower in the area E and similar to the control area (F) as compared to other urban soil samples. This index reached its maximum in samples C and D.

To perform multivariate ordination of microorganism communities we first transformed values of species quantities to the $\chi 2$-metric and calculated the 6 × 6 matrix D of Bray-Curtis distances for all possible pairs in the sampling areas (Table 2).

Table 2. Matrix D of Bray-Curtis distances between microbial communities for all possible pairs in the studied areas. The distances were calculated after the $\chi 2$-transformation of species abundance.

Site	F	A	B	C	D
A	0.737				
D	0.558	0.418			
C	0.717	0.267	0.383		
D	0.781	0.334	0.297	0.278	
R	0.347	0.710	0.609	0.732	0.816

The optimal projection of matrix D on the surface with the NMDS1–NMDS2 coordinate axes was performed by nonmetric multidimensional scaling. Since every test area can be characterized by an abundance of species habitats, there is an opportunity to evaluate projections of ecological optimum points for every specimen on an ordination plot. The latters can be determined as centers of attraction of population density distribution in relation to the studied areas. As in this case the scales both for sampling areas and for species are comparable, it is possible to plot either areas or species on one ordination plane. The reciprocal arrangement of the sampling areas and the composition of soil microorganism species are shown on the combined ordination plot (Fig. 3).

As it is vivid from the plot, the control area F and the area E, both located far from the highway in the park area, have similar numerical compositions of bacteria. The area B can be highlighted as a somewhat peculiar case where heavy metal contamination was minimal. There is a likelihood that this peculiarity caused by an abundance of bacteria originated from the shielding effect of perennial shrubs that grew near the highway and were able to prevent the ingress of traffic pollutants at this site 30 m distanced from the pollution source.

Thus, on the basis of spent ordination the following division of 6 sites into 2 groups is accepted: 1 – with influence of polluting emissions (sites A–D) and 2 – conditionally not polluted sites (background F and site E). Positions of species coordinates on the ordination diagram (Fig. 3) correspond to indicator value $IndVal_j$ calculated for this two groups and presented in the Table 1. In particular, the species with high indices, such as *Mycobacterium* sp., *Streptococcus mutans*, *Acetobacter* sp., *Propionibacterium jensenii*, *P. freudenreichii* etc. positioned to the left by the $NMDS_1$ scale, where the

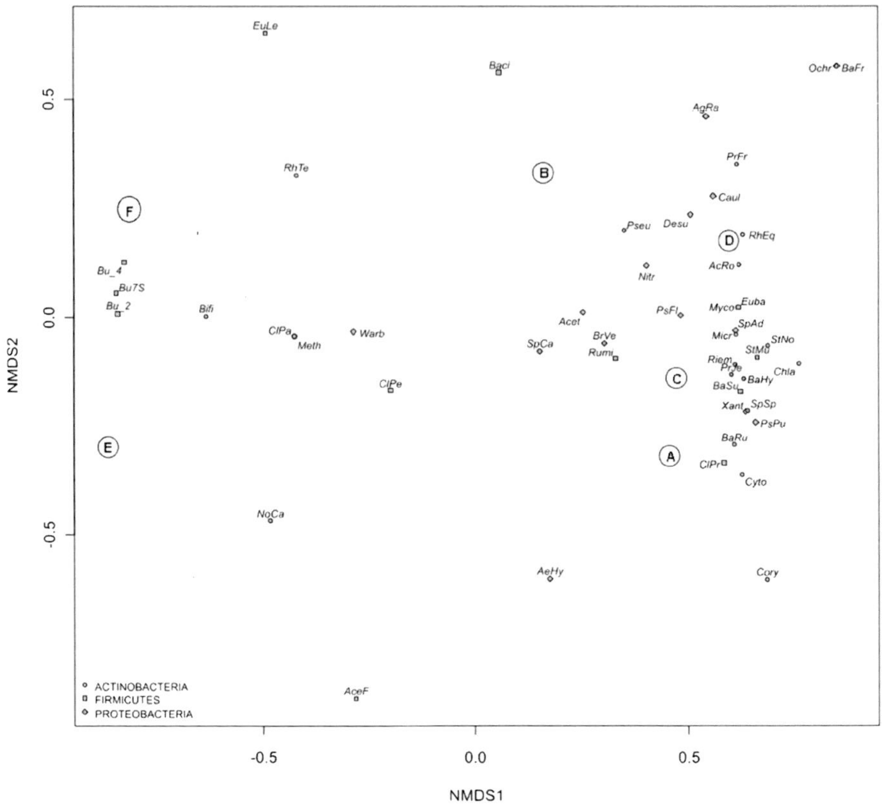

Fig. 3. Ordination of nonmetric multidimensional scaling data: (a) sampling sites (A–F); (b) microbial species (for codes see Table 1).

most polluted sites are situated (group 1). In turn, a cluster of taxonomic groups *Butyrivibrio* and *Bifidobacterium sp.* is an indicator of soils under relatively low level of contamination (group 2). Calculated *p*-values of significance for indicatory indices are shown in the Table 1.

As informativity of traditional bioindicatory indices is high, the results of this study suggest high robustness of overall soil microorganisms abundance index to heavy metals pollution. Indicator significance of actinobacteria was slightly higher than that of micromycetes. Abundance of the micromycetes decreased in soil samples with high level of content of heavy metals.

4 Conclusions

The results of this study allow a conclusion that the chemodiagnostic method based on the use of fatty acids, corresponding aldehydes and the content of hydroxy acids offers very good possibility to identify differences in the structure of urban soil microbiota. The highest indicator value was assigned to *Actinobacteria* phylum, whose content decreased remarkably in polluted samples. Indicator significance of actinobacteria was higher than that of micromycetes. Abundance of the micromycetes decreased in soil samples with high level of content of heavy metals only and anaerobes *Butyrivibrio* sp. and *Bifidobacterium* sp. were regarded as an indicator for soils with relatively low pollution exposure. Therefore, it has a definite bioindicator value not only for sanitation and epidemiology practice but also for expert assessment of the environmental condition.

Acknowledgments. This work was supported by Russian Science Foundation [grant 14-50-00029, in part Microbiological Analysis Using Lipid Markers] and by program of Presidium of RAS # I. 21П Biodiversity of natural systems. Biological resources of Russia: assessment of the state and fundamental bases of monitoring [in part of Soil Chemical Properties].

References

1. Almås, Å.R., Afanou, A.K., Krogstad, T.: Impact of lignosulfonate on solution chemistry and phospholipid fatty acid composition in soils. Pedosphere **24**(3), 308–321 (2014). https://doi.org/10.1016/S1002-0160(14)60017-5
2. Azarbad, H., van Straalen, N.M., Laskowski, R., Nikiel, K., Röling, W., Niklińska, M.: Susceptibility to additional stressors in metal-tolerant soil microbial communities from two pollution gradients. Appl. Soil. Ecol. **98**, 233–242 (2016). https://doi.org/10.1016/j.apsoil.2015.10.020
3. Barreiro, A., Fontúrbel, M.T., Lombao, A., Martín, A., Vega, J.A., Fernández, C., Carballas, T., Díaz-Raviña, M.: Using phospholipid fatty acid and community level physiological profiling techniques to characterize soil microbial communities following an experimental fire and different stabilization treatments. CATENA **135**, 419–429 (2015). https://doi.org/10.1016/j.catena.2014.07.011
4. Davydova, S.: Heavy metals as toxicants in big cities. Microchem. J. **79**(1–2), 133–136 (2005). https://doi.org/10.1016/j.microc.2004.06.010
5. Dufrene, M., Legendre, P.: Species assemblages and indicator species: the need for a flexible asymmetrical approach. Ecol. Monogr. **67**(3), 345–366 (1997)
6. Frostegard, A., Tunlid, A., Baath, E.: Phospholipid fattyacid composition, biomass, and activity of microbial communities from 2 soil types experimentally exposed to different heavy metals. Appl. Environ. Microbiol. **59**(11), 3605–3617 (1993)
7. Frostegård, Å., Tunlid, A., Bååth, E.: Shifts in the structure of soil microbial communities in limed forests as revealed by phospholipids fatty acid analysis. Soil Biol. Biochem. **25**(6), 723–730 (1993). https://doi.org/10.1016/0038-0717(93)90113-P
8. Ge, C.-R., Zhang, Q.-C.: Microbial community structure and enzyme activities in a sequence of copper-polluted soils. Pedosphere **21**(2), 164–169 (2011). https://doi.org/10.1016/S1002-0160(11)60114-8
9. Grimm, N.B., Faeth, S.H., Golubiewski, N.E., Redman, C.L., Wu, J., Bai, X., Briggs, J.M.: Global change and the ecology of cities. Science **319**(5864), 756–760 (2008). https://doi.org/10.1126/science.1150195
10. Gu, Y.-G., Gao, Y.-P., Lin, Q.: Contamination, bioaccessibility and human health risk of heavy metals in exposed-lawn soils from 28 urban parks in southern China's largest city, Guangzhou. Appl. Geochem. **67**, 52–58 (2016). https://doi.org/10.1016/j.apgeochem.2016.02.004

11. Azarbad, H., van Straalen, N.M., Laskowski, R., Nikiel, K., Röling, W.F., Niklińska, M.: Susceptibility to additional stressors in metal-tolerant soil microbial communities from two pollution gradients. Appl. Soil. Ecol. **98**(Febuary), 233–242 (2016). https://doi.org/10.1016/j.apsoil.2015.10.020
12. Kapkiyai, J.J., Karanja, N.K., Qureshi, J.N., Smithson, P.C., Woomer, P.L.: Soil organic matter and nutrient dynamics in a Kenyan nitisol under long-term fertilizer and organic input management. Soil Biol. Biochem. **31**, 1773–1782 (1999)
13. Laczko, E., Rudaz, A., Aragno, M.: Diversity of anthropogenically influenced or disturbed soil microbial communities. In: Insam, H., Rangger, A. (eds.) Microbial Communities, pp. 57–67. Springer, Heidelberg (1997). https://doi.org/10.1007/978-3-642-60694-6_6
14. Osipov, G.A., Turova, E.S.: Studying species composition of microbial communities with the use of gas chromatography-mass spectrometry: microbial community of kaolin. FEMS Microb. Rev. **20**, 437–446 (1997)
15. Shekhovtsova, N.V., Osipov, G.A., Verkhovtseva, N.V., Pevzner, L.A.: Analysis of lipid biomarkers in rocks of the Archean crystalline basement. In: Proceedings of SPIE, vol. 4939, pp. 160–168 (2003)
16. Legendre, P., Gallagher, E.D.: Ecologically meaningful transformations for ordination of species data. Oecologia **129**, 271–280 (2001)
17. Luo, X.-S., Yu, S., Zhu, Y.-G., Li, X.-D.: Trace metal contamination in urban soils of China. Sci. Total Environ. Vol. **421–422**, 17–30 (2012). https://doi.org/10.1016/j.scitotenv.2011.04.020
18. McCune, B., Grace, J.B.: Analysis of Ecological Communities. MjM Software Design, Gleneden Beach (2002)
19. Seto, K.C., Rodńguez, R.Ś., Fragkias, M.: The new geography of contemporary urbanization and the environment. Ann. Rev. Environ. Resour. **35**, 167–194 (2010). https://doi.org/10.1146/annurev-environ-100809-125336
20. Stroganova, M.N., Myagkova, A.D., Prokof'eva, T.V.: The role of soils in urban ecosystems. Eurasian Soil Sci. **30**(1), 82–86 (1997)
21. Tessier, A., Campbell, P.G.G., Bisson, M.: Sequential extraction procedure for the speciation of particulate trace metals. Anal. Chem. **51**(7), 844–851 (1979)
22. Zelles, L., Bai, Q.Y.: Fractionation of fatty acids derived from soil lipids by solid phase extraction and their quantitative analysis by GC-MS. Soil Biol. Biochem. **25**(4), 495–507 (1993). https://doi.org/10.1016/0038-0717(93)90075-M
23. Zenova, G.M.: Actinomycetes in biogeocenoses/soil organisms as biogeocenosis components. Science, 162–170 (1984). (in Russian)
24. Zhang, Q.-C., Wang, G.-H., Yao, H.-Y.: Phospholipid fatty acid patterns of microbial communities in paddy soil under different fertilizer treatments. J. Environ. Sci. **19**(1), 55–59 (2007). https://doi.org/10.1016/S1001-0742(07)60009-6

Soil Functional-Environmental Evaluation and Monitoring in Urban Ecosystems: Principal Functions, Background Objects and Uniform Algorithms of Assessment

Ivan I. Vasenev[1,2(✉)]

[1] Russian Timiryazev State Agrarian University (RTSAU), Moscow, Russia
ivvasenev@gmail.com
[2] Far Eastern Federal University (FEFU), Vladivostok, Russia

Abstract. The paper presents the regional automated system of soil functional-environmental assessment (RASSCA) reflecting soil provincial genesis and functional features. It allows evaluating soil integral functional and environmental quality even in heterogeneous land plots. The RASSCA identifies and quantifies the limiting factors and principal parameters of soils' functional quality and environmental status as well as priority issues of their land-use. The RASSCA conducts environmental computer simulation, search and target prediction of soil quality changes and environmental problems' solution that improve the effectiveness of land-use decision-making – by prior functional-environmental analysis and comparison of the various scenarios of their outcomes and consequences. The RASSCA was used for the functional-environmental analysis of urban soils in the Chernozemic region of Russia (Kursk city), where natural soils are dominated by Chernic Phaeozems and Luvic-Chernozems. We focused on the polluted urban soils to identify the key environmental problems of soil technogenic degradation. The RASSCA allowed evaluating soil limiting parameters and environmental factors to identify the priority issues and to model the target changes in the integrated functional-environmental quality of soils by improving their limiting parameters. Conducted monitoring of Kursk soil technogenic successions highlighted the importance of soil metamorphic and migration processes. The over-compaction in urban soils was up to 0.3 g cm^{-3} $year^{-1}$; an average rate of humification/dehumification – 3.0 g kg^{-1} $year^{-1}$; acidification/alkalization −0.1 pH $year^{-1}$ and heavy metals migration – up to 50–60 g cm m^{-2} $year^{-1}$. Some of these data considerably extend the known ranges of elementary soil processes rates "in situ" and allow for better estimation of the real potentials of soil functioning in different landscape-environmental conditions of forest-steppe zone.

Keywords: Ecology · Soil evaluation · Soil function · Monitoring · Soil assessment algorithm · Soil contamination · Urban ecosystems · Urban soils

1 Introduction

The possibilities of sustainable development in urban areas are crucially limited, often due to the acute and complex environmental conditions. This problem is evident in a variety of modern soil research such as investigations of production-improvement

V.I. Vasenev et al. (eds.), *Megacities 2050: Environmental Consequences of Urbanization*,
Springer Geography, https://doi.org/10.1007/978-3-319-70557-6_17

processes in agriculture and forestry, and developing general solutions to particular environmental problems in urban areas [1–3]. Modern renovation of land ownership [4] and urban conditions [5] in Russian Federation add to the complexity and relevance of both environmental and assessment and monitoring problems. The gap that exists between the research and the applied and practical use of the results is still one of most fundamental problems of the modern environmental soil science fields in Russia for executing effective land monitoring and management.

The urban setting induces a set of specific soil-driving and limiting ecological factors: extreme short-distance spatial variability and dynamics, the age of settlement and complicated land-use history, zoning, human-altered soil profiles, temperature and moisture regimes. Urban soil's health and cover patterns have significant and variable impacts on air and water quality, biodiversity and bio-productivity in urban ecosystem as well as the landscape sustainability and the overall quality of life.

Intense anthropogenic degradation of modern soils increased spatial variability and complexity of soil cover patterns. It's important to develop an adequate tool and regulatory databases for effective evaluation of the environmental roles of soils currently fulfilling economic and environmentally-important functions [6, 7].

The diversity and variation of the size, type and compositions of urban zones results in extreme spatial variability of various soil-environmental functions (SEFs). This is important to recognize and evaluate for urban planning and when developing environmentally-friendly land-use designs. Soil profile variability and distribution generates complex spatial heterogeneity of SEFs [8–10].

The issues of urban soil monitoring refers to the degree of deviation of principal soil diagnostic parameters (PSDP) from their reference or optimum values. In the case of soil functional assessment, the degree of its suitability for environmental and socio-economical performance is considered by mapping a specialized set of SPDP with established criteria and regionally adapted applied evaluation scales [11–13].

A distinctive feature of soil evaluation involves the assessment of the soil's functional quality and ecological status. Current research on the environmental functional assessment of technogenic altered and often contaminated urban soils is under active methodological and normative-methodical developments. The optimal use of such assessments is the practical application by experts of different authorities to determine land use patterns gets possible by developing decision support systems (DSS) with capabilities of automated land evaluation, searches and normative predictions [7–14].

Specialized frameworks for functionally and regionally adapted automated land evaluation systems (ALES) [6, 7, 12] are growing in importance for identifying critical environmental problems and for determining rational versions of urban land-use. This model incorporates the potential for rapid changes in the soil functions by using functional and multiplicative evaluation algorithms-making it a multi-factor regulatory framework.

2 Methodology

The environmental-functions monitoring of urban soil is understood to be organized in time and space as a function of periodic investigation of representative soils' content, structure, regimes and fluxes in their interaction with other components of ecosystems (biota, surface and groundwater, snow, air and so on). Ultimately, this monitoring generates recommendations for the appropriate use of soils in terms of environmental function, while conserving the basic environmental functions and ecosystem services, and minimizing environmental impact risks of urban land-use, and maintaining [4, 11].

The evaluation of the environmental function of soil is a key component of this monitoring. It is an effective tool for identifying and evaluating the critical environmental problems and soil functions, the limiting ecological factors and soil principal diagnostic parameters, and the level of soil improvement in specific urban ecosystems. The quality of such assessment is determined by the reasonable set of soil principal diagnostic parameters – sufficient but not excessive for quantitative analysis, search and normative prediction of concrete types of problematic environmental situations. For each SPDP standard data base is developed with normalized variations of its value from optimal ones varied per different soil subtypes and land-use issues [12, 13].

The issue of local soil standards (soil background object) is especially difficult for the case of urban ecosystems with different anthropogenic impacts and different background soils landscapes. Verification of local soil standards should consider the detailed morphogenetic investigation of soil cover patterns in the reference natural or less disturbed areas with comparative mesorelief, parent material and land-use history. Neglecting these conditions constrains the assessment of urban soils' formation, processes, functions and degradation. Soil environmental zoning (e.g. industrial, residential or recreational) is also relevant for justification of regional and local soil standard. However, especial attention shall be paid on the variability of soil profiles within landscape, subsoils, parent materials texture and other principal features.

Last achievements of soil reference bases development and modern information technologies created good framework conditions for developing regional expert information systems for assessment and modeling ecological state of soils (SAMESs). They are widely used in RF European regions with increased spatial variability and fast development of traditional and new urban areas.

The regional automated system for soil complex assessment (RASSCA) was developed in the RTSAU Lab of agroecological monitoring, ecosystem modeling and prediction (LAMP). The RASSCA system is based on the 4-level algorithms of evaluation (1–4) based on the suggested earlier method of soil quality ecological evaluation [13]. The following principles are considered:

- A particular evaluation of each simple variable quality for uniform soil cover;
- A generalized evaluation of soil quality per its main ecological or environmental functions (using functionally diagnostic groups of soil-environmental parameters);
- An integral quality evaluation of the homogeneous soil cover;
- An integral quality evaluation of the heterogeneous soil cover.

Particular evaluations have been determined from the logistic or model equations, which are valid in the range from critical to optimum values of parameter (1):

$$q_i = (p_{t,i} - p_{k,i}) / ((p_{t,i} - p_{k,i}) + (p_{o,i} - p_{t,i}) exp(-p_i/a_i)) \quad (1)$$

where $p_i = (p_{t,i} - p_{k,i})/(p_{o,i} - p_{t,i})$, a_i- *index of influence of a parameter i,*
$p_{t,i}$, $p_{k,i}$, $p_{o,i}$- *actual (measured), critical and optimum values of a parameter i.*

The RASSCA system was adapted to the natural and man-changed soil types and subtypes that dominate in the investigated urban ecosystems. The main aim (and distinctive feature) of soil environmental assessment is to place its current phase state on the curve (or at the surface) of the soil environmental function and concrete SPDP (Fig. 1). Soil environmental quality is usually determined by soil morphogenetic, physical-chemical, biological parameters and contamination characteristics. The reference of every parameter for logistic equation includes its optimum and critical values, influence index and degradation crisis rate, calculated by RASSCA. The influence index is described as soil quality variety range, which system retains rather stunted changes of its states in. The degradation crisis rate is the maximum of logistic equation simplified derivative for parameters of regional-genetic soil subtype.

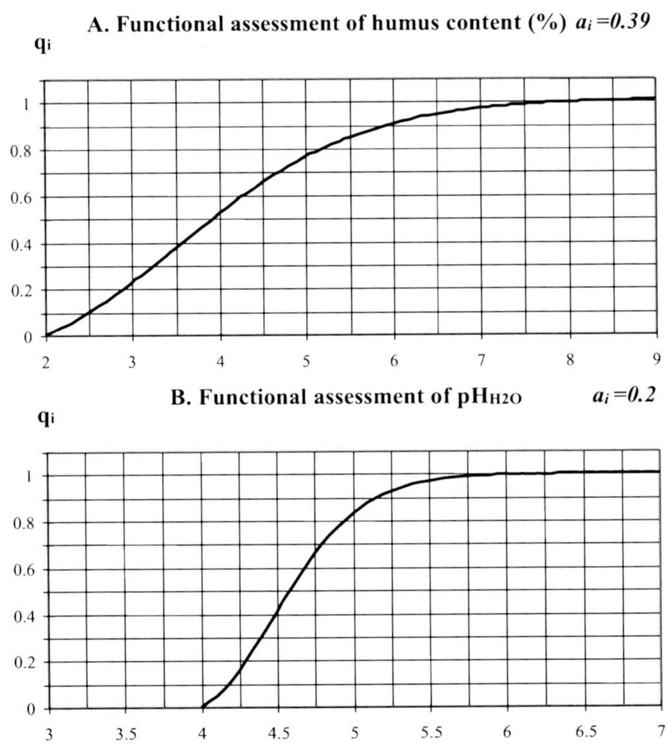

Fig. 1. Graphical presentation of soil humus and pH environmental evaluation for Luvic Chernozems and urban soils in the Central Chernozemic Region of Russia.

Generalized evaluation of quality can be calculated per 3–8 functionally diagnostic groups of parameters. The principal list of them usually includes the following ones: (*a*) soil chemical factor of bio productivity; (b) soil physical factor of plant uptake, soil support for infrastructure; (c) soil spatial heterogeneity factor; (d) soil profile factor of fertility stability and reclaiming potential; (e) soil microclimatic factor; (f) soil phyto sanitary factor; (g) soil contamination factor; (h) soil factor of ecological buffer capacity. Especial attention is taken on soil green house gases (GHG), water and energy fluxes, which are estimated by localized pedotransfer functions or measured in situ [14, 15]. Multiplication functions and harmonic/algebraic mean calculation are useful for generalized (2) and integrated (3) evaluation with universal applied interpretation (Table 1).

Table 1. Interpretation table for soil functional environmental evaluation

Factor	Soil ecological and environmental functions assessments according to their values					
	1-0.96	0.95-0.76	0.75-0.51	0.50-0.26	0.25-0.01	0.0
A.	LAND with high	LAND with increased	LAND with average	LAND with low	LAND with very low	LAND unproductive
	Bio productivity					
B.	very good	good	rather good	bad	very bad	extremely bad
	Support for infrastructure					
C.	homogeneous	weak	semi-	high	unsuitable for use in	
		heterogeneous			grass loans	parks
D.	not requiring	requiring insignificant	requiring significant	requiring	requiring general reclamation	
	Potential for reclamation improving					
E.	very		rather		very	extremely
	favorable			unfavorable		
	Soil microclimatic conditions					
F.	very good	good	rather good	bad	very bad	extremely bad
	Phytosanitary state					
G.		relative	un-	relative	not enough	un-
	satisfactory sanitary-ecological state			suitable for agriculture		
H.	high	increased	average	low	very low	deprived
	resistance to pollution					
Q_Σ	optimal	good	satisfactory	unsatisfactory	very bad	critical
	SOIL INTEGRAL FUNCTIONAL and ENVIRONMENTALSTATE					

$$Q = m\Pi\left(q_i^n\right) / \Sigma\left(\Pi\left(q_i^n\right)/q_i^n\right) \quad (2)$$

where m - number of the factor analyzed parameters; n - parameter stability index.

$$Q = \sqrt[8]{A * B * C * D * E * F * G * H} \quad (3)$$

Generalized and integrated evaluation involves the observable complex of SPDP with special emphasis on limiting parameters and factors. The method assesses the resilience (buffer) indexes of soil ecological state and environmental functions. Soil cover patterns and heterogeneity coefficients are considered for assessment an integral quality evaluation of the heterogeneous soil cover [13].

Most difficult and crucial issues of soil functional environmental assessment are (*1*) to set up the characteristic-factor/quality-requirement subsystems; (*2*) to choose main SPDP for one's partitioning. Usually we used for soil assessment up to 35 primary characteristics and add to principal equations of particular and generalized evaluation specialized simulation models. Special attention is focused on procedures of simulation for GHG fluxes and soluble pollutant migration in the landscape.

Developed system on soil functional environmental evaluation RASSCA reflects their main provincial-genetic features and allows: (1) to give robust assessments of soil environmental state; (2) to reveal and to estimate quantitatively the priority of main ecological (environmental) problems; (3) to carry out computer simulation and predictions of problem situation development; (4) to simplify the process of an administrative solution acceptance in environmental field.

3 Research Area

We tested this framework evaluation system for the case of Kursk city – the administrative center of Kursk region, with biggest population and urban extent in the west part of Central Chernozemic region of Russia. It is located in the central part of the Middle Russian upland, beyond the probable limits of maximum glaciation, within the large-hilly strongly dissected area. The territory is located in the central part of the forest-steppe zone with precipitation/evapotranspiration index around 1 and domination of Haplic and Luvic Chernozems. Kursk-city soils include urban soils and semi-natural Haplic and Luvic Chernozems and Chernic Phaeozems with different intensities of anthropogenic disturbance. The parent material varies from sandy loam to clay silty loams. Deep humus profile, fine structure, favorable physical-chemical properties and high supplies of nutritions characterize the natural soils in the region. Haplic and Luvic Chernozems are often studied in the Central Chernozemic Reserve, located close to Kursk and representing the undisturbed standards of Chernozem formation [16, 17].

Considering worsening of ecological state currently occurring in the Russia Chernozemic regions [4, 18], complex functional environmental investigations are needed to identify critical environmental problems and to develop the most suitable solutions in terms of land-use optimization.

Kursk city has a long history. The research sites were located in the old industrial district with urban soils based on Chernic Phaeozemas and Luvic Chernozems. The reference natural soils of an undisturbed forests and virgin steppes at the 15 km distance from the research sites were monitored for comparative analysis.

4 Results and Discussion

The results obtained urban soils based on the Luvic Chernozem indicated their degradation due to intensive human impacts for last 50 years (Table 2). Soil contamination was the main factor decreasing soil ecological state, however, critical values were also reported for pH_{KCl}. Evaluation modeling showed that improvement of only two soil parameters could already essentially increase the integral quality of this soil after its lime treatment.

Table 2. Functional environmental assessment (FEA) of urban soils based on the Luvic-Chernozem and Chernic-Phaeozemin Kursk

Soil pit #	Soil ecological (environmental) factor (SEF)	Factor FEA	Limiting SPDP/factor	SPDP value	SPDPFEA	Factor FEA after improvement of limiting factor	
						1-st	2-nd
Urban soils based on the Luvic Chernozem							
11	Soil contamination factor	0.46	$Pb_{mob.}$, ppm	56	0.22	0.81	0.91
	Soil factor of ecological buffer capacity	0.73	pH_{KCl}	4.9	0.41	0.95	0.99
	Soil profile factor of fertility stability & reclamative state	0.71	pH_{KCl}	4.9	0.41	0.96	0.99
	Soil chemical factor of bio productivity	0.71	pH_{KCl}	4.9	0.41	0.96	0.99
	Soil physical factor of plant uptake and soil workability	0.99	No			1.0	
	Integrated evaluation	0.70	Soil contamination factor			0.93	0.98
Urban soils based on the Chernic Phaeozem							
12	Soil contamination factor	0.90	$Pb_{mob.}$, ppm	38	0.76	0.90	0.92
	Soil factor of ecological buffer capacity	0.33	Bulk density, g cm^{-3}	1.55	0.14	0.56	0.82
	Soil profile factor of fertility stability & reclamative state	0.38	Bulk density, g cm^{-3}	1.55	0.14	0.51	0.93
	Soil chemical factor of bio productivity	0.56	Humus, %	3.06	0.24	0.56	0.93
	Soil physical factor of plant uptake and soil workability	0.41	Bulk density, g cm^{-3}	1.55	0.14	0.79	0.98
	Integrated evaluation	0.55	Soil physical factor			0.65	0.93

The results obtained urban soils based on the Chernic-Phaeozemin the old industrial zone, indicated decreasing of soil buffer capacity as the principal limiting factor. However, a negative influence of over-compaction was also essential. Evaluation modeling showed that even improvement of three limiting soil parameters could not essentially increase the integral quality of this soil. In this case, especial attention shall be given to the 2-nd level limiting parameters and multi-version modeling of their potential improvement consequences on the soil quality.

The outcomes obtained for urban soils based on the Chernic-Phaeozem are typical for the analysis of soil buffer capacity in case of contaminated urban soils (Table 3). In case of urban soils exposed to recreation degradation and chemogenic transformations, soil acidity dominated the limiting factors. Therefore, liming could solve the principal problem situation and improve the two other limiting SPDP (i.e CEC and its saturation) essentially. As for the urban soils, based on the Chernic-Phaeozemthe principal problem situation could be solved only after essential increasing of 2-nd limiting parameter - humus content and 3-rd limiting parameter - CEC. The costs of soil reclamation for the second case were substantially higher.

Table 3. Ecological buffer capacity assessment and modelling of urban soils based on the Luvic-Chernozem and Chernic-Phaeozem in Kursk: assessment and modelling results

Soil pit #	SPDP	Value	Limit #	FEA	Predicted	
					Value	FEA
Urban soils based on the Luvic Chernozem						
11	Clay content, %	44	No	0.99		
	Humus content, %	5.64	3	0.87	6.5	1.0
	CEC, mEq kg^{-1}	193	4	0.94	200	0.95
	CEC saturation, %	70	2	0.65	95	1.0
	pH_{KCl}	4.9	1	0.41	6.5	1.0
	Bulk density, g cm^{-3}	1.07	No	1.0		
	Ecological buffer capacity			0.73	0.99 (*1, 2, 4*), 1 (*1–4*)	
Urban soils based on the Chernic Phaeozem						
12	Clay content, %	42	No	0.98		
	Humus content, %	3.06	2	0.24	5.0	0.85
	CEC, mEq kg^{-1}	174	3	0.78	200	0.95
	CEC saturation, %	94	No	0.99		
	pH_{KCl}	6.5	No	1.0		
	Bulk density, g cm^{-3}	1.55	1	0.14	1.25–1.15	1
	Ecological buffer capacity			0.33	0.56 (*1*), 0.90 (*1–3*)	

The factor of ecological buffer capacity was the most important in our study of the contaminated soils (see Table 1). This factor was limited by the pH and bulk density. Increasing these limitation factors is challenging especially in soils with sandy and sandy-loam texture.

The application of the RASSCA tools in the ecological monitoring of industrially polluted city garden lands in the neighborhood of the Kursk Tannery has resulted in almost 10 times reduction of the total area of the initially planned output of land-use – due to detailed investigation of strong spatial differentiation of soil chromium contamination and their resistance to pollution. A similar analysis of the other city garden soil contamination with copper, nickel and lead also showed the statistically significant negative correlation of their mobility indexes with humus content and pH values, identifying limiting SPDP of soil health and resistance to heavy metals contamination.

Urban impact monitoring in Kursk has shown (Table 4) wide ranges of variation in rates of soil migration and accumulation processes in case of their various anthropogenic transformations with significant intensification of these processes for soils with low ecological buffer capacity.

Table 4. Rates range of accelerated soil processes in urban soils, based on the Luvic-Chernozem and Chernic-Phaeozem in Kursk

Soil processes	SPDP	Rates range
Litter degradation	*Litter stock*–kg m^{-2} $year^{-1}$	up to 3–5
Over-compaction	*Density* –g cm^{-3} $year^{-1}$	up to 0.3
Dehumification	*Humus*–g kg^{-1} $year^{-1}$	1–3
Humification	*Humus*–g kg^{-1} $year^{-1}$	0.6–2
Acidification	*pH*	0.05–0.1
Alkalization	*pH*	0.02–0.03
Carbonisation	*Carbonates*–g kg^{-1} $year^{-1}$	1–1.3
Leaching	*Carbonates*–kg cm m^{-2} $year^{-1}$	1–3
HM contamination	*MAC* $year^{-1}$	up to 0.5–1
HM migration	*Pb, Cr, Cu* – g cm m^{-2} $year^{-1}$	20–30 (60)

Substantial pollution with heavy metals resulted in severe alteration of properties and function of Urban soils based on Chernic Phaeozems and Luvic Chernozems. The irreversible changes in soil functions could result in transformation to the new subtype of urban soils – Chemogenic Urbosoils. These Chemogenic Urbosoils are formed under intensive long-term (20–30 years) technogenic atmospheric impact sand reveal a clearly diagnosed heavy metals migration (Pb, Cd, Ni, Zn, Cu), The depth of this migration can be at 1 m or more, even in the case of loam soil with high humus content. The results of the average annual migration rate research in Kursk city garden lands showed the maximum migration in urban soils based on Luvic-Chernozem- 20–30 g cm m^{-2} $year^{-1}$ for lead and 50–60 g cm m^{-2} $year^{-1}$ – for chromium. The verified models of HM migration allow predicting the future soil environmental quality in case of different level industrial impacts and to diagnose in advance the principal environmental problems in soil regulation function disturbances.

5 Conclusions

The paper presents the regional automated system of soil functional-environmental assessment (RASSCA). Implementing the RASSCA system (i) reflects soils' provincial genetic and functional features; (ii) allows evaluating the integral functional and environmental quality of soils even in conditions of heterogeneous land plots; (iii) gives the effective tools to identify and quantify the limiting factors and principal parameters of the functional quality and environmental status of soils and priority issues of their land use; (iv) provides the useful platform to conduct environmental computer simulation, search and target prediction of soil quality further changes and problem environmental situations solution; (v) helps to improve the effectiveness of land-use and land conservation decision making – by prior functional-environmental analysis and comparison of the various scenarios.

The functional-environmental analysis of polluted urban soils based on Chernic-Phaeozem and Luvic-Chernozem stressed the environmental problems of soil technogenic degradation in Kursk city and neighborhoods. Adapting the RASSCA system to the environmental conditions and factors of the research area allowed to evaluate soil limiting parameters and environmental factors, to identify the priority issues that require urgent improvement, and to model the target changes in the integrated functional-environmental quality of soils by improving the priority set of its limiting parameters.

Conducted monitoring research of Kursk soil technogenic successions showed strong activation of soil typomorphic metamorphic and migration processes in urban soils. The over-compaction in urban soils was up to 0.3 g cm^{-3} $year^{-1}$; an average rate of humification/dehumification – 3.0 g kg^{-1} $year^{-1}$; acidification/alkalization −0.1 pH $year^{-1}$ and heavy metals migration – up to 50–60 g cm m^{-2} $year^{-1}$. Some of these data considerably extend the known ranges of elementary soil processes rates "in situ" and allow for better estimation of the real potentials of soil functioning in different landscape-environmental conditions of forest-steppe zone.

A substantial increase in HM migration rates in urban soils has been observed. The main-driving factors included sandy loam texture, acidification and sharp decreasing of humus content in the profiles without geochemical (lithological, acid-base or redox) barriers. The obtained results can be used for current assessment and forecasting of environmental pollution problem development in urban areas with soil characteristics similar to those investigated in Kursk city.

Acknowledgments. This work was supported by the RF President grant# NSh-10347. 2013.11.

References

1. Addiscott, T.M., Wagenet, R.J.: Concepts of solute leaching in soils: a review of modelling approaches. J. Soil Sci. **36**, 411–424 (1985)
2. Bouma, J.: Land evaluation for landscape units. In: Sumner, M.E. (ed.) Handbook of Soil Science, pp. E-393–E-412. CRC Press, Boca Raton (2000)
3. Sumner, M.E. (ed.): Handbook of Soil Science. CRC Press, Boca Raton (2000)

4. Tchernogorov, A.L., Tchekmaryov, P.A., Vasenev, I.I., Gogmachadze, G.D.: Land Agroecological Evaluation and Land-Use Optimization, 268 p. Moscow University Press, Moscow (2012). (in Russian)
5. Vasenev, V.I., Stoorvogel, J.J., Ananyeva, N.D., Ivashchenko, K.V., Sarzhanov, D.A., Epikhina, A.S., Vasenev, I.I., Valentini, R.: Quantifying spatial-temporal variability of carbon stocks and fluxes in urban soils: from local monitoring to regional modeling. In: Muthu, S.S. (ed.) Carbon Footprint Handbook, pp. 185–221. CRC Press, Boca Raton (2015)
6. Rossiter, D.G.: A theoretical framework for land evaluation. Geoderma **72**, 165–190 (1996)
7. van Lanen, H.A.J., Hack-ten Broeke, M.J.D., Bouma, J., de Groot, W.J.M.: A mixed qualitative/quantitative land evaluation methodology. Geoderma **55**, 37–54 (1992)
8. Prokof'eva, T.V., Gerasimova, M.I., Bezuglova, O.S., Bakhmatova, K.A., Gol'eva, A.A., Gorbov, S.N., Zharikova, E.A., Matinyan, N.N., Nakvasina, E.N., Sivtseva, N.E.: Inclusion of soils and soil-like bodies of urban territories into the Russian soil classification system. Eur. Soil Sci. **47**, 959–967 (2014)
9. Salvati, L., Ferrara, C., Ranalli, F.: Changes at the fringe: soil quality and environmental vulnerability during intense urban expansion. Eur. Soil Sci. **47**, 1068–1075 (2014)
10. Greinert, A.: The heterogeneity of urban soils in the light of their properties. J. Soils Sed. **15**, 1725–1737 (2015)
11. Vasenev, I.I., Buzilev, A.V., Kurbatova, J.A., Chistotin, M.V., Rudnev, N.I., Tiunov, A.I.: Agroecological modelling and design. In: Vasenev, I.I. (ed.) 264 p. RTSAU, Moscow (2010). (in Russian)
12. Vasenev, I.I., Buzilev, A.V.: Land Agroecological Evaluation Automated Systems, 172 p. RTSAU, Moscow (2010). (in Russian)
13. Vasenev, I.I.: Soil and land-use planning. In: Nortcliff, S. (ed.) Task Force: Soil Matters – Solutions Under Foots, pp. 100–104. GeoEcology Essays, Catena Verlag (2015)
14. Wagenet, R.J., Hutson, J.L.: Soil quality and its dependence on dynamic physical processes. J. Environ. Qual. **26**, 41–48 (1997)
15. Vasenev, I.I., Vizirskaya, M.M., Vasenev, V.I., Valentini, R., Raskatova, T.V.: Comparative analysis of principal factors of spatial-temporal variability of CO2 emission from Moscow urban soils with various levels of anthropogenic impact. Izvestia Timiryazev Agric. Acad. (Special Issue), 11–23 (2012)
16. Mikhilova, E.A.: Botanical composition, soil and forage quality under different management regimes in Russian grasslands. In: Mikhilova, E.A., Bryant, R.B., Post, C.J., Cherney, D.J.R., Vassenev, I.I.: Agriculture, Ecosystems and Environment, vol. 80, no. 3, pp. 213–226 (2000)
17. Mikhilova, E.A., Bryant, R.B., Vassenev, I.I., Schwager, S.J., Post, C.J.: Cultivation effects on soil organic carbon and total nitrogen at depth in the Russian Chernozem. Soil Sci. Soc. Am. J. **64**, 738–745 (2000)
18. Sarzhanov, D.A., Vasenev, V.I., Sotnikova, Y.L., Tembo, A., Vasenev, I.I., Valentini, R.: Short-term dynamics and spatial heterogeneity of CO2 emission from the soils of natural and urban ecosystems in the Central Chernozemic Region. Eur. Soil Sci. **48**(4), 416–424 (2015)

Modern Trends of Sustainable Housing Design Using Landscape Urbanism Principles

E. Krasilnikova(✉) and L. Popova

Department of Urbanism and Theory of Architecture,
Institute of Architecture and Urban Development,
Volgograd State Technical University, Volgograd, Russia
landurbanizm@gmail.com, lorik057@gmail.com

Abstract. In terms of the urban planning cities regeneration the post-industrial brownfield landscapes, directly adjacent to the modern residential quarters, become most vulnerable in the structure of urban planning tissue. These "space erosion holes of urban planning fabric" create centers of socially insecure and marginal urban environment formation. The aim of this work is to analyze of modern trends, identification of priority principles of landscape urbanism and its implementation into the formation process of residential structures within these areas to ensure sustainable, socially integrated and environmentally comfortable living environment.

Keywords: Landscape urbanism · Agriculture landscape · Social scenario · Sustainable development

1 Introduction

Currently, the processes affecting globalization increase in the European countries have direct impact on the variation and extension of the approaches, and also changing the opinion of urban planners, architects, politicians and the residents according the creation of different types of socially-oriented housing, adapted to the specific, for each European country, socio-economic conditions within the current world crisis.

In order to understand what kind of the current trends of residential housing design will be the most effective for implementation of urban planning regeneration aims - such as post-industrial brownfield areas - it is necessary to apply innovative teaching methods based on the international cooperation in the field of the students of architectural schools training. This cooperation is very important for students to acquire the skills necessary for their professional interaction with all social and state institutions during their future professional activities (in this aspect the OIKONET example is revealing) [23].

"Opening the city in the nature" by regeneration of post-industrial and brownfield areas, based on the principles of landscape urbanism, is contained in the works of Waldheim [1, 2], Corner [3], Mostafavi and Najle [4], and Krasilnikova in Russia [5], the connection between the city and the natural landscape is forming, and, consequently, the problem of the environmental and micro-climatic characteristics of adjacent areas

V.I. Vasenev et al. (eds.), *Megacities 2050: Environmental Consequences of Urbanization*, Springer Geography, https://doi.org/10.1007/978-3-319-70557-6_18

improving is solving [18]. The relevance of this research is in the fact that the aspect of residential structures formation, from the point of view of the theoretical concept of landscape urbanism, has not been considered yet.

On the basis of the urban planning regeneration of the industrial area project in the Soviet district of Volgograd made first as course work (5th study year), and then the thesis work (6th study year) at the Department of Urbanism and Architecture Theory in Volgograd State Architectural and Civil Engineering University (VSUACE), the method of flexible, stable and socially oriented residential and public environment was offered.

2 The Value of Landscape Urbanism for the Residential Housing Formation (Residential Structures) in Terms of Urban Planning Regeneration

2.1 Background for the Landscape Urbanism Introduction into the Contemporary Processes of the Urban Planning Regeneration

Features of the urban planning regeneration in terms of postmodernism [6] is to strengthen of the migration processes, which leads to the population increase in major European cities. As consequence, the city expands its territorial boundaries by the construction of new residential areas, but not always this process goes at the same with the formation of accessible transport infrastructure and the creation of public landscapes and developed system of social facilities. The occurrence of this problem leads to the deterioration of social and environmental conditions for living on these sites [16].

The negative aspect, from the point of view of balanced urban development, is the territorial expansion of the cities, not only by increasing the number of depressed areas, but also due to the construction of new residential areas for migrants both in Europe and Russia, that usually have low level of site improvement. And this fact results in the deterioration of environmental conditions for living there.

What is the meaning of landscape urbanism for the cities development and reconstruction in terms of the economic crisis and intensifying of the migration processes relating to world globalization? It's wrong to consider Landscape Urbanism only as the mechanism addressing to the environmental problems of the city. There is only one application of landscape urbanism in this aspect. The ideology of landscape urbanism is based on the creation of socially-oriented and environmentally friendly urban environment, and the understanding of the city as the landscape, or as the unified ecosystem [3, 5] includes not only increase of the natural landscape components in the city structure, but also, and the most importantly, the formation identical social-oriented spaces of various functional filling, based on such natural landscape components.

One of the most effective methods for solving this social and urban planning problem is the creation of the landscape and urban planning framework of the city which is an integrated system of interconnected green transport and pedestrian links, public open spaces and the existing natural framework. Formation of the landscape and urban planning framework is based on the scenario approach (which is one of the theoretical basis of landscape urbanism). Considering the fact that the framework has a great potential for continuous development within the space-time conditions, it is expediently to define it as the landscape and urban planning scenario of the city development, Fig. 1.

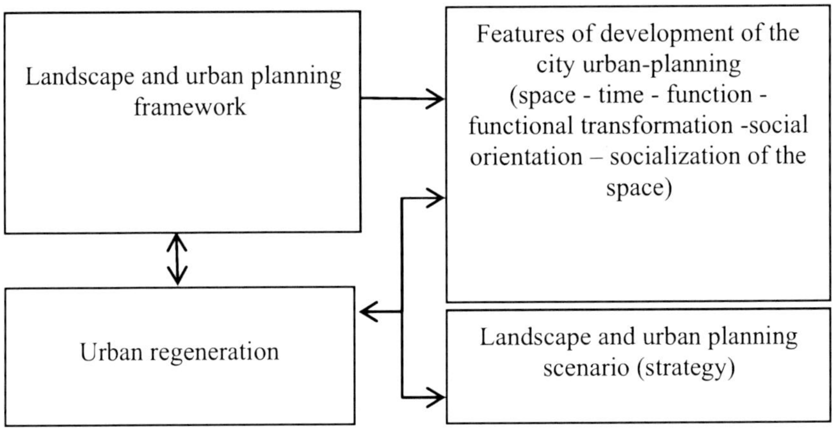

Fig. 1. Integration of landscape and urban planning scenario into the process of urban planning regeneration.

Today this aspect is extremely important for the cities urban planning development and urban renewal, from the standpoint of the comfortable, environmentally friendly urban environment formation that is accessible for all the categories of citizens. Spaces socialization - the formation of the social scenario of the space, and then, based on this process, creation of the landscape and urban planning scenario [20] – creates new urban fabric structure where we see new hybrid spaces, formed on the basis of the architectural and landscape complexes of various functional purpose [19]. Thus, there are new types of residential areas appearing in the city structure, that can usefully be measured as the architectural and landscape complexes.

Currently, various aspects of the creation of the landscape and urban planning scenario of the city development based on the concept of landscape urbanism, are devoted the works of Weller [7, 8], Desimini [9, 31], Reed [10], Kahn [11], Wall [12]. Issues affecting the necessity and urgency of the urban planning cities regeneration strategy from the position of the integrated socio-oriented development, aimed at comfortable living environment creation, are the basis for the following government documents, essential for the territorial development of Europe and Russia.

These documents are - the strategy "Europe 2020" [22], the agreement "European Green Capital" [23], Urban Development Concept Berlin 2030 [24], Urban Planning doctrine of the Russian Federation [25], Summary of the strategy of sustainable development of Russia [26], as well as policy documents of the European Union, such as The Sixth Environment Action Program of the European community [27], Thematic Strategy on the Urban Environment [28], The Leipzig Charter on Sustainable European Cities [29], The Charter of European Sustainable Cities and Towns Towards Sustainability [30].

The main positions of these documents emphasize the importance of the social - oriented, environmentally sustainable and comfortable environment formation for the residents of the city at all its of territorial and object levels (district – center - quarter - residential complex – backyard - house).

2.2 Residential Structures Formation on the Basis of the Residential Architectural and Landscape Complexes Creation

Contemporary trend in the development of scientific and practical concept of landscape urbanism is an active application of the landscape urbanism principles within the process of the urban planning regeneration of European cities. The example of urban planning regeneration of Milan based on the concept of "Green Milan rays" by Kipar [5, 13] is very revealing.

Based on this concept, the redevelopment of industrial areas and inefficiently used areas are carried out. These areas are redesigned not only into new public parks, but residential complexes, developed by world-renowned architects such as Hadid, Libeskind, Isodzaki, Boery, Kipar [13].

The emergence in Europe of the first landscape urbanism objects, that became its "icon", is a good tradition for theory and practice formation of this direction. The European practice of landscape urbanism has direct impact on new opportunities and vectors of its development. The examples of the post-industrial areas regeneration are most illustrative, but the scenarios of development and regeneration of built-up areas with a mixed building are no less interesting. Such an example of modern using of practice landscape urbanism is the regeneration of the 13th district of Paris, the district Rive Gauche, based on the principles of sustainability, reflected in the Environment Development Charter of this district [5].

The formation of architectural and landscape complex Massena quarter (Quartier Massena, architect. Christian de Portzamparc) is based on the maximum inclusion of natural components in its structure, namely - natural style created landscape. Important environmental activities within the formation of the landscape infrastructure of this architectural and landscape complex are:

1. interconnected system of gardening objects with different functionality (park - square - garden - pedestrian promenade - sport and playgrounds - embankments), this approach ensures the maximum permeability of building with the elements of infrastructure landscape;
2. using of advanced environmental technologies in order to create sustainable plantations (picking, treatment and reuse of rainwater and wastewater for irrigation; biological methods instead of chemical in farming practices, etc.);

3. using of drip irrigation (for economical water use during the irrigation);
4. using of secondary ecologically certified materials for infrastructure development;
5. using of energy-saving technologies for the complex maintenance.

Thus, we can see the priority of ecological approach, which is one of the basic principles of the theoretical concept of landscape urbanism. Consequently, the formation of comfortable and environmentally sustainable living environment is aimed at the improving of living conditions and is one of the basic conditions for the housing formation in terms of urban planning regeneration.

Therefore, we can speak about the socialization of space, and the fact that while using the scenario approach we are formatting social-oriented landscape urban planning scenario for the territory, where residential architectural and landscape complexes are the structure's element along with the public spaces, Fig. 2.

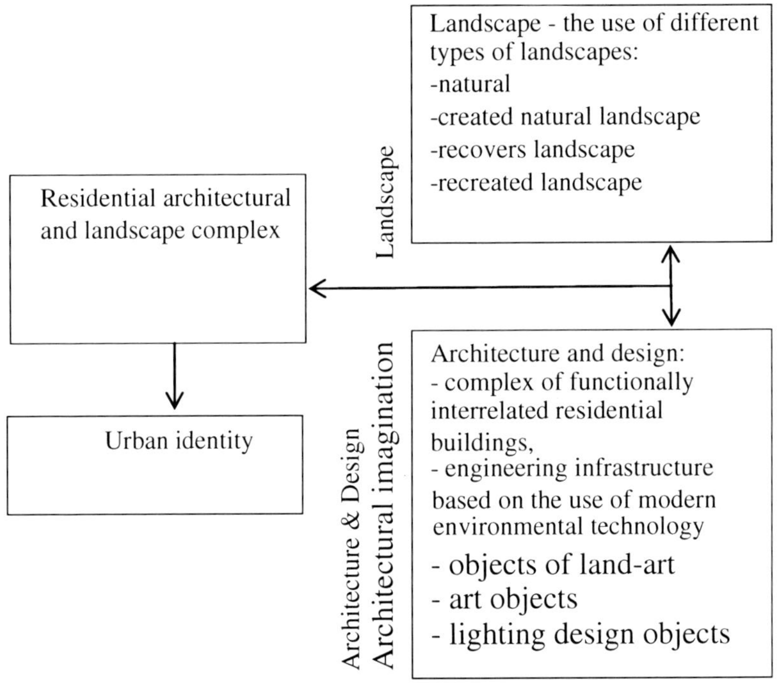

Fig. 2. Formation of residential architectural and landscape complex

The aim of this work is the analyze of modern trends, identification of priority principles of landscape urbanism and its implementation into the formation process of residential structures within these areas to ensure sustainable, socially integrated and environmentally comfortable living environment.

3 The Introduction of the Study of the Residential Structures Formation in the Context of Landscape Urbanism into the Learning Process

3.1 Application of the Landscape Urbanism Principles within the Formation of the Architectural and Landscaped Residential Complexes

Special attention should be paid to the study of the evolution and formation of residential architectural and landscape complexes. The lecture course "Architectural and landscape complexes" has been the part of the basic variable bloc of the program discipline for 4th year Bachelor Degree in Volgograd State University of Architecture and Civil Engineering since 2014, approved by the Ministry of Education of Russia [17]. In this course, the students are researching and analyzing the functional, typological and morphological features of residential architectural and landscape complexes, studying international and Russian counterparts within the urban planning, architectural and landscape analysis of the selected complex.

Application of the landscape urbanism principles in the creation of world-known residential complexes is studied in the frame of students' independent work, and also during the workshops.

For example, the application of such principle of landscape urbanism as horizontally — the priority of working with horizontal surfaces, rather than with vertical structures, characterized by many residential complexes by ZahaHadid. Not all the landscape urbanism principles are applicable in the creation of residential buildings. The analysis of contemporary examples of residential structures formation allows identify three principles that have been successfully used in the design practice in Russia, Europe, America, Singapore and Australia.

These are: the principle of the scenario approach; ecological principle - the adaptation of the design solutions to the environmental conditions and modern environmental technologies use; the principle of horizontality - priority of working with horizontal surfaces and the use of the maximum capacity of the earth surface; the principle of the process formation - programming of residential areas and socio-economic processes development that affects its formation and functioning [1, 2, 5].

3.2 Integration into the International Educational Process

In the frame of grant « OIKOnet. A global multidisciplinary network on housing research and learning » 539369- LLP-1-2013-1-ES-ERASMUS-ENW (at the Department of Urbanism and Theory of Architecture in the Institute of Architecture and Urban Planning Development of VSUACE (chief Prof. Krasilnikova E.E, docent Rusanov V.A.) cooperating with professor V. Joklova (FA STU) during the course work of the 4th year students "Reconstruction of the city's part" (bachelors and specialists) the project of the residential area of Petrzalka (Bratislava, Slovakia) reconstruction on the basis of the use of modern principles of landscape urbanism has been executed [14, 15, 21].

There were three diploma project on urban planning reconstruction of the residential area in Petrzalka (Bratislava) developed, which task was the conceptual design of sustainable urban planning and architectural design model of urban planning renovation of these city areas and introducing there the sustainable and socially-oriented residential units.

Thus, the study of contemporary trends of the sustainable residential structures formation is an important aspect not only for understanding of contemporary development processes of urban planning and space-planning city structure [32], but also for the implementation of the study into the scientific and pedagogical process of teaching within the Architecture specialty.

3.3 New Methods and Approaches Determinating the Main Directions of the Development of the Residential Formation in the Urban Planning Regeneration Conditions

It's expediently to introduce the searching for new methods and approaches of the residential structures design into the graduate design practice at the architectural faculties of universities.

During the past several years the diploma projects on urban planning regeneration of the most problematic territories for the city development has been preparing at the Department of Urbanism and Theory of Architecture in the Institute of Architecture and Urban Planning Development of VSUACE. The diploma project of Manko D. is one of them.

The design object is the typical Volgograd's industrial and residential areas of the Soviet district. Today this site is a spontaneously built up market place with large surrounding areas of unsettled, degrading landscape.

In order to improve the environmental situation and to ensure the territory aeration comfort it's supposed to create "green corridors" reestablishing the connection with the territories remote from the Volga river through the network of landscape and urban planning nodes.

Thus, "green framework" creates the basis for the efficient and environmentally sustainable landscape and urban planning system, within which new residential neighborhoods are designed, Fig. 3.

The concept of residential structures is aimed at the creation of the diverse typology of housing and infrastructure, and its inclusion into the system of public spaces and green areas. Due to the diversity and variability of housing typology the division of private and public spaces is formed. Variability is expressed with the diversity of heights, typology, quarters amount, facades design. The quarters are free from cars that ensure the environment safety. The concentration of public functions within the quarters allows the resident to minimize the personal vehicles use.

Formation of the agriculture landscapes system is presented with gardens and orchards within the residential buildings and public spaces structure. The residents of the surrounding neighborhoods can easily use it by growing agriculture products for their own consumption or for sale. Thus, agriculture landscapes accumulate not only the design territory, but also the surrounding neighborhoods and quarters. They fill up

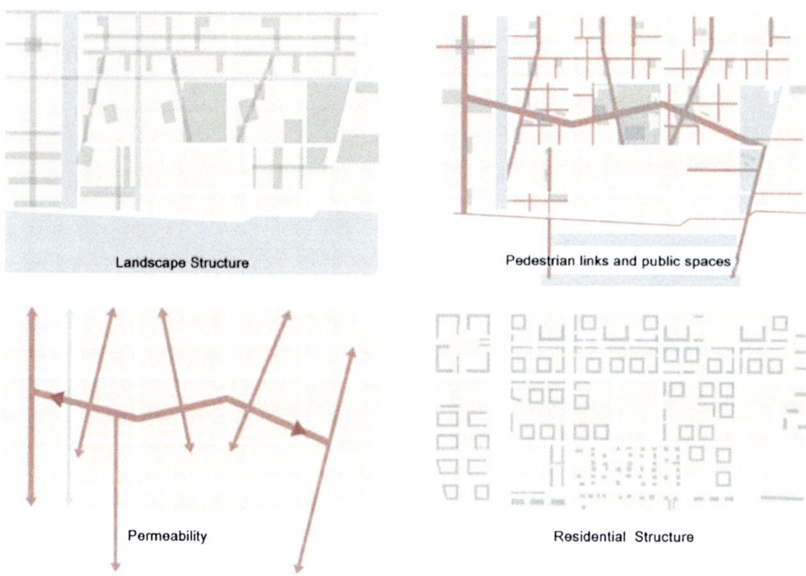

Fig. 3. The concept of the sustainable residential structure creating

the citizens'everyday life with new social and economic functions, and can be included into the city daily life scenario. Considering that the greening system includes the agriculture landscapes, there is need for its integration into the social scenario of citizens' life. The main pedestrian dominant, curved in plan, stretches the entire width of the design area and provides permeability of spaces.

In addition to the collective gardens and season kitchen gardens adjacent to the houses areas and public facilities there will be all-season facilities arranged, Fig. 4. These structures are based on the agriculture landscapes round-year use - greenhouses, glasshouses, klimatrons.

The concept of the new way of living can be described as the daily filling up of social environment with functions and providing the platforms for their implementation and development. In this context, it is pertinently to create "urban agricultures" - small areas of urban space given over to agricultural use and economically connected with the city. New forms of agriculture emerged in the city and its surrounding areas, have different meanings: it is not just food production but also fuel costs reducing, while the multicultural approach during the process contributes to the unification of the different age and social groups.

The cities plots of land for food growing can be really found even there where it seems impossible. Recreating of the mixed use areas may become a global challenge.

Thus, we can talk not only about formation of the new residential urban planning system based on the landscape urbanism principle, but also about the formation of the new model of the city life based on the cooperation and joint consumption, Fig. 5.

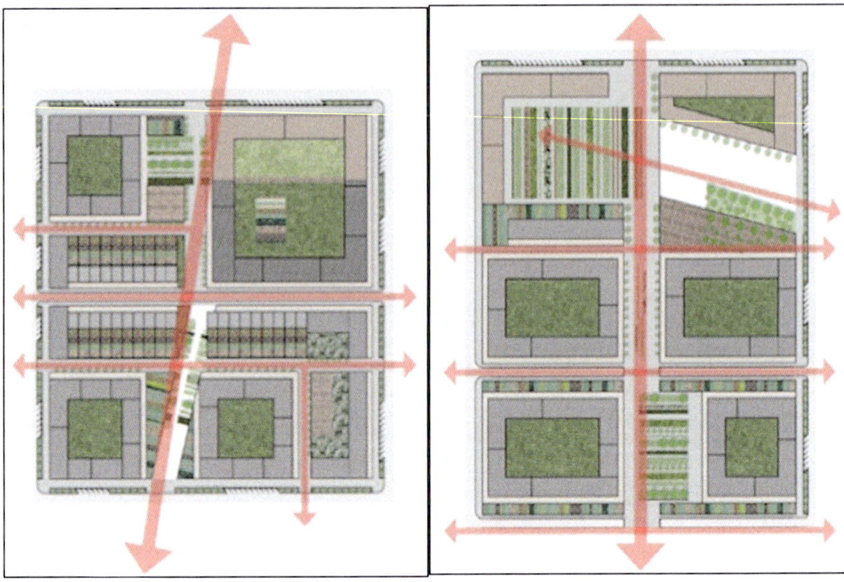

Fig. 4. Permeability of the quarters

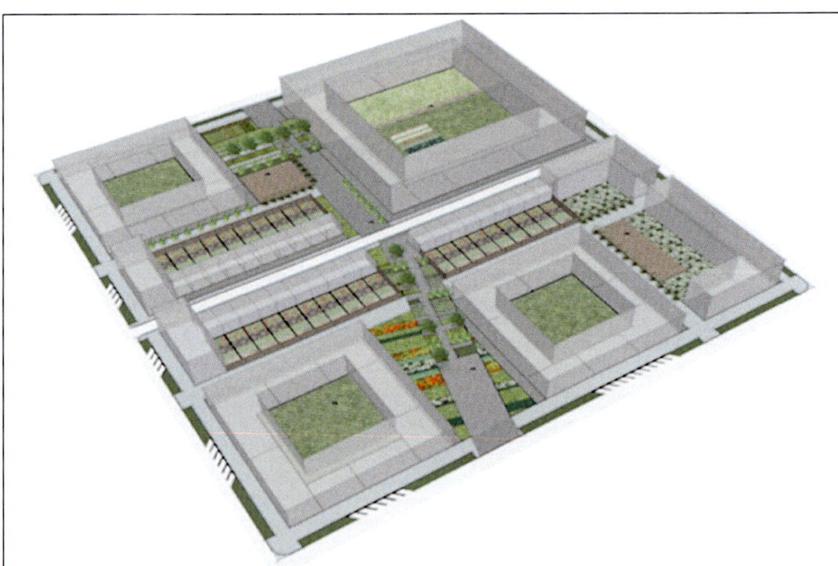

Fig. 5. The model of the quarter formed with the principles of landscape urbanism

4 General Conclusions and Reflection

Modern trends in the development of new residential building types based on the landscape urbanism principles allow creating new types of social relationships at the level of local urban communities.

Interdisciplinary identity of landscape urbanism does not limit the possibility of creating the unique urban areas on different levels, including the architectural and residential landscapes.

Maximum inclusion of natural components into the residential buildings structure and implementation of the natural components capacity for changing during the growing season allows using the changeable dynamics of the visual perception for the identical image of the residential development. This is an important aspect within the modern model formation of the sustainable housing.

References

1. Waldheim, C.: Landscape as Urbanism A General Theory. Princeton University Press Princeton and Oxford, New Jersey (2016)
2. Waldheim, C. (ed.): The Landscape Urbanism Reader. Princeton Architectural Press, New York (2006). p. 43
3. Corner, J., Bick Hirsch, A.: The Landscape Imagination, Collected Essays of James Corner 1990–2010. Princeton Architectural Press, New York (2014)
4. Mostafavi, M., Najle, C.: Landscape Urbanism: A Manual for the Mechanic Landscape Architectural Association, London (2003)
5. Krasilnikova, E.: Landscape Urbanism. Theory-Practice: Part.1: Scientific and Practical Foundations of Landscape Urbanism, LTD IAA « District news », Volgograd (2015)
6. Ellin, N.: Integral Urbanism. Routledge, New York (2006)
7. Weller, R.: Landscape (Sub) Urbanism in Theory and Practice. Landscape J. **27**(2), 255–278 (2008). Board of Regents of the University of Wisconsin System
8. Weller, R.: Planning by Design - Landscape architectural scenarios for a rapidly growing city. Journal of Landscape Architecture. In: ECLAS (ed.), JoLA, autumn 2008, 18–29, Munich, pp. 6–17 (2008)
9. Desimini, J.: Blue voids: storm water strategies for abandoned lands. Journal of Landscape Architecture In: ECLAS (ed.), JoLA, autumn 2013, 08-13-2013. Munich: Callwey, pp. 64–73, 2013
10. Reed, C.: Landscape Urbanism in Practice, Topos, pp. 90–107, June 2010
11. Kahn, E.A.: Scenario city, COA: Central Office of Architecture, On Landscape Urbanism, Center 14, Austin TX: Center for American Architecture and Design University of Texas at Austin School of Architecture, pp. 196–203 (2008)
12. Wall, A., Boomtown v. Regiocity: Thinking and designing for the networked cityregion, on Landscape Urbanism, Center 14, Center for American Architecture and Design University of Texas at Austin School of Architecture, Austin TX, pp. 304–313 (2008)
13. Troshina, M.: Milan waiting, Project International 38, Joint projects, pp. 18–105 (2014)
14. Krasilnikova, E., Rusanov, V., Kuzina, L.: Resilient and sustainable housing - contemporary aspects of forming. Bulletin of Volgograd State University of Architecture and Civil Engineering Series: Civil Engineering and Architecture, vol. 36, no. 55, pp. 275–283 (2014)

15. Mrđenović, T., Devetaković, M., Jolkova, V., Krasilnikova, E.: IPI methodology for designing resilience habitat regeneration strategies using OIKONET. In: Changing Cities II Spatial, Design, Landscape & Socio-economic dimension Proceedings, edAspaGospodini. University of Thessaly, Greece, pp. 770–782 (2015)
16. Hadjri, K., Durosayie, I.O., Csanadj, G., Czismady, A., Olt, G., Devetakovic, M., Mrdjenovic, T., Joklova, V., Madrazo, L., Krasilnikova, E., Kuzina, L.: A critical analysis of urban regeneration programmes. In: Europe, Conference: Housing-a critical perspective, Architecture MPS, Liverpool University. Liverpool John Moores University, Liverpool, 08–09 April, pp. 1–18 (2015)
17. Krasilnikova, E.: The working program of educational discipline B.3.V.06. Architectural and landscape complex towards: 270100 Architecture 270300 Design of architectural environment and profile training (Bachelor): Architectural Design Urban Design, Architecture and constructive design, design of the urban environment, interior design. Publishing house VSUACE, Volgograd (2012)
18. Krasilnikova, E.: Landscape urbanism a new look at an old problem. J. Green City **4**(11), 31–35 (2014)
19. Krasilnikova, E., Klimov, D.: Role and value of landscape urbanism in the modern process of city development and reconstruction. In: Balkan Architectural Bienale, Conference Proceedings, "capital a", BAB, Belgrade, Serbia. http://www.balkanarchitecturalbiennale.rs/documents/BAB_Conference_Proceedings_15_prot.pdf
20. Krasilnikova, E.: Landscape and urban planning transformation of space-planning structure. The Hybrid Link #03 Hybridization between Form and Energy, pp. 1–26. http://www.urbanhybridization.net
21. OIKOnet: http://oikonet.org/
22. Europe 2020: http://ec.europa.eu/europe2020/index_en.htm, http://eur-lex.europa.eu/LexUriServ/LexUriServ.do?uri=COM:2010:2020:FIN:EN:PDF
23. European Green Capital: http://ec.europa.eu/environment/europeangreencapital/index_en.htm
24. Urban development concept Berlin 2030: http://www.stadtentwicklung.berlin.de/planen/stadtentwicklungskonzept/index_en.shtml
25. Urban planning doctrine of Russian Federation: http://uar74.ru/news/documents/115
26. Summary of the strategy of sustainable development of Russia: http://www-sbras.nsc.ru/win/sbras/bef/strat.html
27. The sixth environment action programme of the European community 2002–2012. http://ec.europa.eu/environment/archives/eap/index.htm
28. Thematic strategy on the urban environment: http://eur-lex.europa.eu/legal-content/EN/TXT/?uri=uriserv%3Al28171
29. The Leipzig Charter on Sustainable European Cities: http://www.espon-usespon.eu/library,leipzig-charter-on-sustainable-european-cities
30. The Charter of European Sustainable Cities and Towns Towards Sustainability, Aalborg. http://www.sustainablecities.eu/aalborg-process/charter
31. Desimini, J.: Wild innovation: Stoss in Detroit. Scenario J. (2013). http://scenariojournal.com/article/wild-innovation-stoss-in-detroit/
32. References Krasilnikova E urban regeneration Dublin. http://www.oikodomos.org/workspaces/app/webroot/files/references/text/vjoklova_24_prof._E.Krasilnikova_Urban_regeneration__Dublin__30_oct.2014.pdf

Author Index

V.I. Vasenev et al. (eds.), *Megacities 2050: Environmental Consequences of Urbanization*, Springer Geography, https://doi.org/10.1007/978-3-319-70557-6

CPI Antony Rowe
Chippenham, UK
2019-03-13 04:07